Advances in Polymer Science

Fortschritte der Hochpolymeren-Forschung

Volume 14

Edited by

H.-J. Cantow, Freiburg i. Br. · G. Dall'Asta, Milano · J. D. Ferry,
Madison · H. Fujita, Osaka · M. Gordon, Colchester · W. Kern, Mainz
G. Natta, Milano · S. Okamura, Kyoto · C. G. Overberger, Ann Arbor
W. Prins, Syracuse · G. V. Schulz, Mainz · W. P. Slichter, Murray Hill
A. J. Staverman, Leiden · J. K. Stille, Iowa City · H. A. Stuart, Mainz

With 11 Figures

Springer-Verlag Berlin · Heidelberg · New York 1974

Editors

Prof. Dr. H.-J. Cantow, Institut für Makromolekulare Chemie der Universität, 7800 Freiburg i. Br., Stefan-Meier-Str. 31, BRD

Prof. Dr. G. Dall'Asta, Istituto di Chimica, Industriale del Politecnico, Piazza Leonardo da Vinci 32, Milano, Italia

Prof. Dr. J. D. Ferry, Department of Chemistry, The University of Wisconsin, Madison 6, Wisconsin 53706, USA

Prof. Dr. H. Fujita, Osaka University, Department of Polymer Science, Toyonaka, Osaka, Japan

Prof. Dr. M. Gordon, University of Essex, Department of Chemistry, Wivenhoe Park, Colchester C04 3SQ, England

Prof. Dr. W. Kern, Institut für Organische Chemie der Universität, 6500 Mainz, BRD

Prof. Dr. G. Natta, Istituto di Chimica Industriale del Politecnico, Milano, Italia

Prof. Dr. S. Okamura, Department of Polymer Chemistry, Kyoto University, Kyoto, Japan

Prof. Dr. C. G. Overberger, The University of Michigan, Department of Chemistry, Ann. Arbor, Michigan 48104, USA

Prof. Dr. W. Prins, Department of Chemistry, Syracuse University, Syracuse, N.Y. 13210, USA

Prof. Dr. G. V. Schulz, Institut für Physikalische Chemie der Universität, 6500 Mainz, BRD

Dr. William P. Slichter, Bell Telephone Laboratories Incorporated, Chemical Physics Research Department, Murray Hill, New Jersey 07971, USA

Prof. Dr. A. J. Staverman, Chem. Laboratoria der Rijks-Universiteit, afd. Fysische Chemie I, Wassenaarseweg, Postbus 75, Leiden, Nederland

Prof. Dr. J. K. Stille, University of Iowa, Department of Chemistry, Iowa City, USA

Prof. Dr. H. A. Stuart, Institut für Physikalische Chemie der Universität, 6500 Mainz, BRD

ISBN 3-540-06649-7 Springer-Verlag Berlin · Heidelberg · New York
ISBN 0-387-06649-7 Springer-Verlag New York · Heidelberg · Berlin

The use of general descrive names, trade marks, etc. in this publication, even if the former are not especially identified, is not to be taken as a sign that such names, as understood by the Trade Marks and Merchandise Marks Act, may accordingly be used freely by anyone.

This work is subject to copyright. All rights are reserved, whether the whole or part of the material is concerned, specifically those of translation, reprinting, re-use of illustrations, broadcasting, reproduction by photocopying, machine or similar means, and storage in data banks. Under § 54 of the German Copyright Law where copies are made for other than private use, a fee is payable to the publisher, the amount to the fee to be determined by agreement with the publisher. © by Springer-Verlag Berlin · Heidelberg 1974. Library of Congress Catalog Card Number 61-642. Printed in Germany. Typesetting and printing: Brühlsche Universitätsdruckerei, Gießen

Contents

Correlation between Cationic Model and Polymerization Reactions of Olefins
    J. P. Kennedy and S. Rengachary 1

Photoinitation of Vinyl Polymerization by Aromatic Carbonyl Compounds
    J. Hutchison and A. Ledwith 49

Chemical Transformations of Cellulose
    L. S. Gal'braikh and Z. A. Rogovin 87

Correlation between Cationic Model and Polymerization Reactions of Olefins

J. P. KENNEDY and S. RENGACHARY

Institute of Polymer Science, The University of Akron; Akron, Ohio 44325, USA

Table of Contents

I. Introduction	2
II. A Critical Review of Model Studies in Cationic Polymerization	3
III. Experimental Part	13
A. Materials	13
B. Procedure	13
C. Analyses	16
IV. Results and Discussion	17
A 1. The Validity of the Kennedy-Gillham Scheme	17
A 2. Extension of the Kennedy-Gillham Scheme to Other Alkylaluminum Compounds	20
A 3. Parallelism Between Model and Polymerization Reactions	22
B. Model Studies	23
B 1. Reactivity of Alkylaluminums	26
B 2. Coinitiator Efficiency of Alkylaluminums	27
B 3. Effect of Initiator Systems on Transfer (Elimination) and Termination (Alkylation) A New Termination Mechanism	29
B 4. Effect of Alkylaluminums on Propagation	33
B 5. Effect of Solvents	34
B 6. Comparison of Alkylaluminum-Alkyl Halide Initiator Systems in Methyl Halide Solvents	35
B 7. Conclusions from Model Experiments	36
C. Polymerization Studies	38
C 1. Comparison Between Model and Polymerization Studies	45
V. Summary	46
VI. References	48

I. Introduction

The purpose of this study was to investigate the mechanism of cationic olefin polymerizations by model experiments using alkylaluminum/alkyl halide initiator systems and to correlate the results of model experiments with corresponding polymerization reactions.

Simple organic substances and oligomers have often been used to elucidate reaction mechanisms, reactivities of functional groups, and structural characteristics of polymers. However, only a limited amount of model studies have been carried out in the field of cationic polymerization.

In recent years certain alkylaluminum compounds e.g., Et_2AlCl, Me_3Al, in conjunction with suitable proton, halonium and carbenium ion sources (e.g., HCl, Cl_2, t-BuCl) have proved to be effective initiator systems for producing high molecular weight polymers from olefins, dienes, vinyl and cyclic ethers (1–5). In contrast to conventional Lewis acids e.g., $AlCl_3$, $AlBr_3$, BF_3, the above initiator systems yield high molecular weight polymers at relatively high temperatures and also are relatively insensitive to traces of moisture or other impurities. The above discovery also led to model studies employing alkylaluminum coinitiators by Kennedy and coworkers (6, 7) and Priola and coworkers (8). The former group studied under various conditions, the reaction between t-butyl halides and Me_3Al yielding neopentane, whereas Priola and coworkers investigated the reaction between t-BuCl and ethylaluminum derivatives. While these model studies led to a better understanding of the influence of solvents and ethylene in these reactions, they have not been directly correlated with polymerizations.

The first part of this paper is a critical review of model studies in cationic polymerization. In the second part we describe and discuss our investigations exploring the effect of a variety of alkylaluminum/alkyl halide initiating systems under a variety of conditions on the competitive reactions of the Kennedy-Gillham scheme (9). This scheme represents a comprehensive set of model reactions developed for the study of competitive reactions in cationic olefin polymerization. It involves the cationation of a nonpolymerizable (steric-hindrance) olefin under simulated polymerization conditions and the complete analysis of reaction products which in turn reflect initiation, propagation, termination and transfer.

In addition to a better understanding of the effect of reaction variables on products, rates and molecular weights, this work has also led to new mechanistic concepts such as termination by hydridation and to the optimization of cationic graft-copolymer synthesis.

II. A Critical Review of Model Studies in Cationic Polymerization

In a historically important paper Whitmore (*10*) outlined the mechanism of cationic polymerization by determining the structures of products formed in the sulfuric acid induced oligomerization of isobutylene. Despite the long history of cationic polymerization, the fact that acid catalyzed olefin polymerizations involve carbenium ions was proposed only in 1934 by Whitmore. This author gave a full description of initiation, propagation, termination and transfer reactions and recognized their role in the polymerization of olefins by protonic acids. The first step is the addition of a proton to the double bond resulting in a positively charged carbon (initiation) which, depending on conditions undergoes: a) combination with a negative ion (termination), b) reversal of the initiation process by the loss of the same or a different proton to give the same or a different olefin (transfer and/or unimolecular termination) and c) addition to another olefin resulting in a larger positive species (propagation), which can again undergo all the above mentioned reactions. Whitmore substantiated this theory by analyzing the structure of the products i.e., dimers, trimers and tetramers of isobutylene. Whitmore's scheme, however, does not include a true kinetic termination step and the products formed by initiation and transfer cannot be distinguished. The use of a labelled initiator e.g., D_2SO_4, could prove informative towards this end.

Recently two ponderous papers appeared by Kriz and Marek which purported to be model studies of the cationic polymerization of isobutylene (*11, 12*). In the first phase of their study (*11*) the authors chose 1,1'-dineopentyl ethylene, a trimer of isobutylene, as their model compound and worked with $AlBr_3$, HBr (or DBr) in n-heptane at $-70°$ (*11*). They mixed these reagents by the following sequence: nC_7 + HBr + trimer + $AlBr_3$. The products were analyzed by gas chromatography, NMR (60 MHz), and mass spectroscopy. Useful, hard data of this research are as follows: 1. 1,1'-dineopentylethylene rapidly isomerizes to 2,2',4,6,6'-pentamethyl-3-heptene:

$$\text{(structure)} \xrightarrow[C_7, -70°]{AlBr_3/HBr} \text{(structure)} + \text{other isomers}.$$

2. The reaction is essentially over in a few seconds. 3. In case DBr is used, multiple deuterium incorporation occurs, the position analysis of which is extremely difficult. 4. Slow isomerization occurs in the presence of $AlBr_3$, *alone*, in the absence of HBr or H_2O.

Against a backdrop of our understanding of carbenium ion reactions, only the last finding is genuinely new. Unfortunately, exact reaction conditions have not been published and the details of a direct isomerization experiment in which only 1,1'-dineopentylethylene and $AlBr_3$ was used has not been presented so that the data cannot be sufficiently scrutinized. But accepting the conclusions of the Czechoslovak authors on face value, it is evident that carbenium ions (or at least carbenium-ion like species) may form from certain olefins in the presence of strong Lewis acids alone even at low temperatures. The kinetics of the isomerization of 1,1'-dineopentylethylene induced by $AlBr_3$ (and other strong Lewis acids) should be reinvestigated as it might provide clues to the details of initiation of cationic polymerization of olefins in the absence of protogenic or cationogenic substances.

Another word of caution in respect to this research. As mentioned, the addition sequence of the reagents was: olefin + HBr + $AlBr_3$. This mixing sequence might give rise to alkyl bromides which could subsequently react with $AlBr_3$, obscuring the results. A better mixing sequence would be: olefin + $AlBr_3$ + HBr. A sample should be withdrawn after the addition of the second component; the analysis of this would reveal the status of the system prior to addition of the last ingredient.

In their subsequent paper (*12*) the Czechoslovak authors extended their research over other isobutylene trimers and investigated the reaction which occurs in the system: individual isobutylene trimers + HBr (or DBr) + $AlBr_3$ in n-heptane at $-70°$. The following trimers have been used:

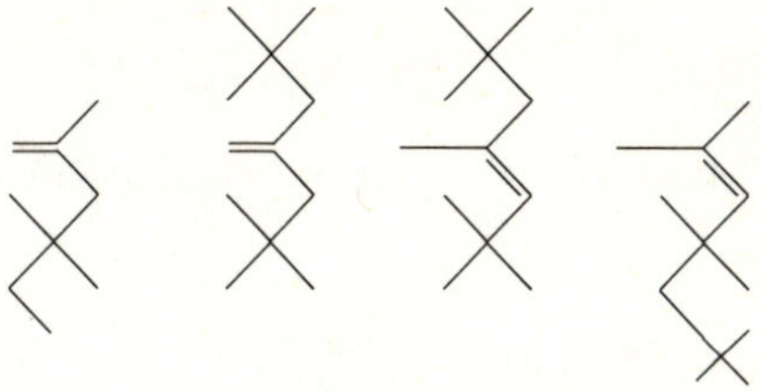

                    I   II   III   IV

Structure III can exist in two geometrically different (a cisoid and a transoid) configurations, however, the exact nature of the isomer has not been specified.

The main findings can be paraphrased as follows: 1. Positional (double bond migration) and skeletal isomerizations occur with all trimers. 2. In addition to multiple isomerizations, all trimers give rise to some unidentified tetramers, pentamers, and hexamers, the most important contribution coming from the pentamers. 3. Extensive deuterium exchange occurs between olefins.

Again the results are reasonable in view of known carbenium ion mechanisms. An important conclusion is that after protonation these highly branched olefins are prone to crack and the smaller carbenium ion fragments then are able to cationate the larger olefins leading to higher olefins. For example:

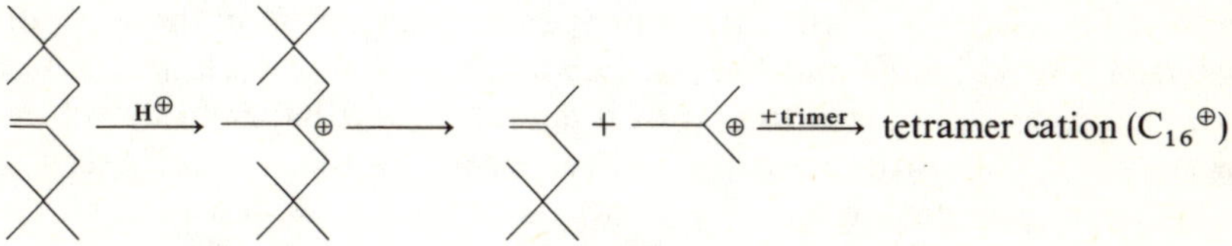

$$\xrightarrow{H^\oplus} \quad \oplus \longrightarrow \quad + \quad \oplus \xrightarrow{+\text{trimer}} \text{tetramer cation } (C_{16}^\oplus)$$

All in all these studies have not shed much new light on the mechanistic details of cationic polymerizations.

Subsequent to his discovery that alkylaluminum compounds in conjunction with suitable alkyl halides are effective initiator systems for synthesizing high molecular weight polyisobutylene (1), Kennedy in 1970 (6) studied the reaction between alkyl and aryl halides with trimethylaluminum using methyl chloride solvent at $-78°$ C. The results were explained in terms of the following reaction scheme:

$$R-Cl + Me_3Al \xrightleftharpoons{MeCl} [R^\oplus \ Me_3AlCl^\ominus] \qquad (1)$$

$$[R^\oplus \ Me_3AlCl^\ominus] \longrightarrow R-Me + Me_2AlCl. \qquad (2)$$

Thus Eq. (1) represents the dissociation of alkyl (aryl) halide by Me_3Al and the formation of alkyl (aryl) cation and counter anion. Equation (2) shows the alkylation of the cation by methide anion from the counter anion and the formation of a hydrocarbon.

The major findings of this study are: 1. The reactivity of the alkyl halides follows the order $3° > 2° > 1°$ which corresponds with the known ease of formation and the stability of the carbenium ions formed. Indeed the methylation by Me_3Al of tertiary alkyl halides emerged as a new synthetic procedure for the preparation of quaternary methyl groups, a difficult task in synthetic organic chemistry. 2. Isobutyl chloride reacts with Me_3Al to yield neopentane, suggesting the presence of carbenium ion intermediates i.e.,

$$iBu^\oplus \longrightarrow t\text{-}Bu^\oplus \xrightarrow{+Me_3AlCl^\ominus} (CH_3)_4C.$$

The above mechanism [Eqs. (1) and (2)] has been viewed by Kennedy as a model for cationic initiation immediately followed by termination.

Kennedy and coworkers (7) further studied the effect of solvent (MeCl, MeBr, MeI and cyclopentane) and the nature of the halogen (Cl, Br or I) in the t-butyl halide on the rate of neopentane formation in the reaction between t-butyl halide and Me_3Al. The findings of this work are: 1. The rate of neopentane formation with different t-butyl halides follows the order $t\text{-BuCl} > t\text{-BuBr} > t\text{-BuI}$. 2. The nature of the solvent exerts a significant influence on the rate of alkylation, i.e., rate decreases in the order $MeCl > MeBr > MeI \gg$ cyclopentane. 3. The rate of neopentane formation is first order in $[t\text{-BuX}]$ and in $[Me_3Al]$. 4. The activation energies in the range $-20°$ to $-80°$ are ~ 11 kcal/mole for all the methyl halide solvents and ~ 16 kcal/mole for cyclopentane.

Trimethylaluminum exists as a dimer in solid, liquid and vapor phase and in nonpolar solvents. However, the active species is the monomeric form. The methyl halide solvent helps to break up the methyl bridge of the Me_3Al dimer enabling the t-butyl halide to interact with the alkylaluminum to generate the initiating $t\text{-Bu}^\oplus$ ion (13). Kinetic analysis led the authors to propose that the rate determining step is the displacement by t-BuCl of the methyl halide from the $Me_3Al \cdot ClMe$ complex:

$$(Me_3Al) \leftarrow ClMe \underset{\text{slow}}{\overset{(CH_3)_3CCl}{\rightleftarrows}} Me_3Al \leftarrow ClC(CH_3)_3 \qquad (3)$$
$$\overset{\text{fast}}{\rightleftarrows} (CH_3)_3C^\oplus \; Me_3AlCl^\ominus .$$

In the presence of a cationically polymerizable monomer the transitory carbenium ion may initiate polymerization. The above study provides insight into the nature of the initiating species and it emphasizes the active role played by solvents which are normally considered to be inert. The slow rates observed in cyclopentane could be due to the absence of a suitable agent that can function as a "bridge opener" for alkylaluminum dimers. Investigations of polymerizations of isobutylene using the above initiators in MeCl, MeBr, MeI and cyclopentane solvents with trimethylaluminum support the conclusions of model studies (14). Thus, the efficiency of polymerization decreases in the order MeCl > MeBr ≫ MeI and cyclopentane. Essentially no polymerization was observed in MeI and cyclopentane.

. Using a similar technique Priola and coworkers (8) studied the reaction between t-butyl chloride and Et_3Al, Et_2AlCl, $EtAlCl_2$ and $AlCl_3$ using methylene chloride and methyl chloride solvents at $-78°$ C, for 2 h. The results of this study can be summarized as follows: 1. The major reaction products are isobutane and 2,2-dimethylbutane in reactions where Et_3Al is used. 2. Other products such as 2,3-dimethylbutane, isooctane and 1-chloro-3,3'-dimethylbutane are also formed in amounts strictly dependent on the molar ratios of t-BuCl/alkylaluminum or the chlorine content of the alkylaluminum. 3. When Et_2AlCl or $EtAlCl_2$ react with t-BuCl, the product consists of branched C_4, C_6, C_8 hydrocarbons and a higher alkyl chloride. 4. Interestingly, $AlCl_3$ does not react with t-BuCl, however, it yields a crystalline complex at $-78°$ in the absence of an added olefin.

These authors interpreted their results in terms of a cationic mechanism involving the following reactions: 1. The carbenium ion abstracts a β-hydrogen from an ethyl group linked to aluminum with the simultaneous formation of ethylene. 2. The carbenium ion adds to ethylene. 3. The carbenium ion abstracts an ethyl group or a chloride ion from the counter anion.

Extensions of these conclusions to polymerization should be made with care. The amount of t-BuCl used in Priola et al.'s experiments was much greater than what is required for polymerization i.e., t-BuCl/AlR$_3$

>1 and the reaction time of 2 h was much longer than ncecessary for polymerization. Moreover, the yield of highly branched products increased both by increasing the t-BuCl concentration and decreasing the ethyl content (or increasing the chlorine content) of the alkylaluminum compound. Thus, the possibility that some of the reaction products obtained with Et_3Al or Et_2AlCl arise in "post reactions" induced by the higher halogenated alkylaluminums i.e., $EtAlCl_2$ or $AlCl_3$

$$Et_3Al + t\text{-BuCl} \rightleftarrows [t\text{-Bu}^\oplus Et_3AlCl^\ominus] \rightarrow t\text{-BuEt} + Et_2AlCl$$
$$t\text{-BuCl} + Et_2AlCl \rightarrow t\text{-BuEt} + EtAlCl_2 \text{ etc.}$$
(4)

cannot be ruled out. The observation that NMR spectroscopy does not show the presence of any t-BuCl even when used in excess tend to support this view. Further, the proposed addition of ethylene to the carbenium ions cannot be significant in polymerizations, where the ethylaluminums are used only in catalytic amounts. The formation of $C_6H_{13}Cl$ via

$$Me_3CCl + Et_2AlCl \rightleftarrows Me_3C^\oplus Et_2AlCl_2^\ominus \quad (5)$$

$$Me_3C^\oplus Et_2AlCl_2^\ominus \rightarrow Me_3CH + C_2H_4 + EtAlCl_2 \quad (6)$$

$$H_2C{=}CH_2 \hookrightarrow Me_3C\text{--}CH_2\text{--}CH_2\text{--}Cl + Et_2AlCl \quad (7)$$

has not been proven. Considering the excess t-BuCl in the system and the weaker bond strength of C–Cl as compared to Al–Cl, it is possible that the chlorine end group in $C_6H_{13}Cl$ is due to transfer between the carbenium ion and t-BuCl.

Optically active compounds have been employed to investigate the nature of intermediates in model cationic reactions involving alkylaluminum compounds. Alberola (15) in 1969 investigated the stereochemistry of alkylation using $(-)\alpha$-phenylethyl chloride with Et_3Al in p-xylene. The formation of predominantly racemic product was taken as evidence for carbenium ion-counter anion intermediates. Kennedy, Desai and Sivaram (7) modified this scheme to avoid undesirable side reactions by using a nonreactive solvent such as ethyl chloride at $-65°$ C. Analysis of the products again confirmed the fact that the stereochemical consequence of alkylation is predominant racemization which is consistent with the hypothesis that the intermediate involved in alkylation is a loose carbenium ion-counteranion pair.

In 1970 Kennedy and Gillham (9) proposed a comprehensive scheme for the study of competitive reactions in cationic polymerization of isobutylene. Their scheme was based on reacting an initiator-coinitiator system with a nonpolymerizing olefin. In particular, they examined the t-BuCl/Me$_3$Al system with 2,4,4-trimethyl-1-pentene, the model compound for isobutylene, using methyl chloride at $-78°$ C. The overall reaction scheme is shown in Fig. 1.

According to this scheme, the reaction between t-BuCl and Me$_3$Al generates t-butyl cation (tBu$^\oplus$) [Eq. (8)], which t-butylates 2,4,4-trimethyl-1-pentene (C$_8^=$) [Eq. (9)]

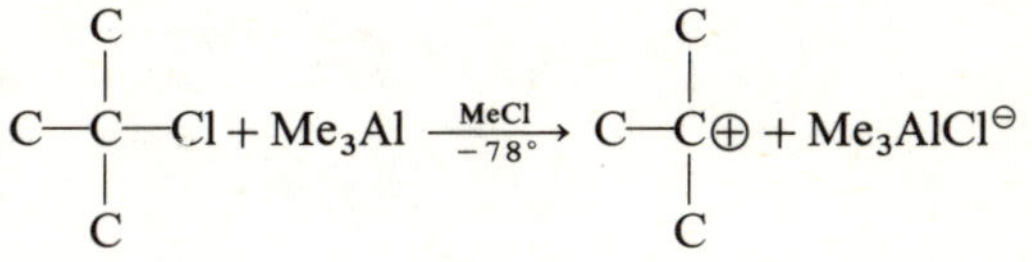

$$\underset{(t\text{-BuCl})}{\text{C–C(C)(C)–Cl}} + \text{Me}_3\text{Al} \xrightarrow[-78°]{\text{MeCl}} \underset{(t\text{-Bu}^\oplus)}{\text{C–C(C)(C)}^\oplus} + \text{Me}_3\text{AlCl}^\ominus \qquad (8)$$

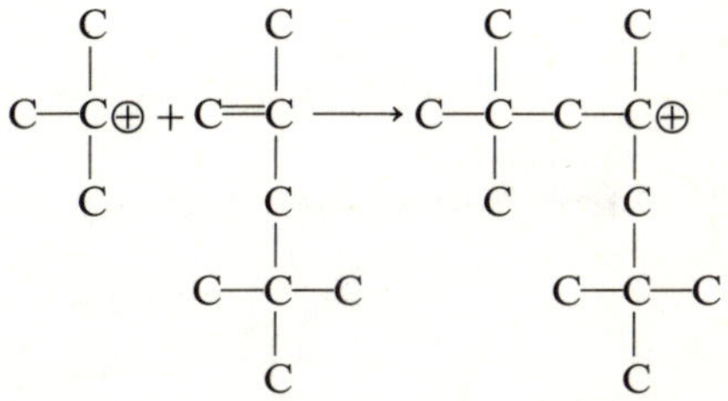

$$\underset{(t\text{-Bu}^\oplus)}{\text{C–C(C)(C)}^\oplus} + \underset{(C_8^=)}{\text{C=C–CH}_2\text{–C(C)(C)C}} \longrightarrow \underset{(C_{12}^\oplus)}{\text{C–C(C)(C)–C–C}^\oplus\text{(C)–CH}_2\text{–C(C)(C)C}} \qquad (9)$$

(for simplicity, the hydrogens on carbon and the counteranion Me$_3$AlCl$^\ominus$ are omitted). The t-butylation of C$_8^=$ olefin to give the C$_{12}^\oplus$ ion is equivalent to initiation. Subsequently the C$_{12}^\oplus$ can undergo: a) addition to another C$_8^=$ molecule forming C$_{20}^\oplus$ [propagation, Eq. (10)], b) methylation by methide transfer from the counteranion Me$_3$AlCl$^\ominus$ [termination, Eq. (11), and c) elimination of a proton giving C$_{12}^=$ (chain transfer, Eq. (12))]:

$$C_{12}^\oplus + C_8^= \rightarrow C_{20}^\oplus \qquad (10)$$

$$C_{12}^\oplus + \text{Me}_3\text{AlCl}^\ominus \rightarrow C_{12}\text{Me } (C_{13}) \qquad (11)$$

$$C_{12}^\oplus + \text{Me}_3\text{AlCl}^\ominus \rightarrow C_{12}^= + [\text{H}^\oplus \text{Me}_3\text{AlCl}^\ominus] \qquad (12)$$

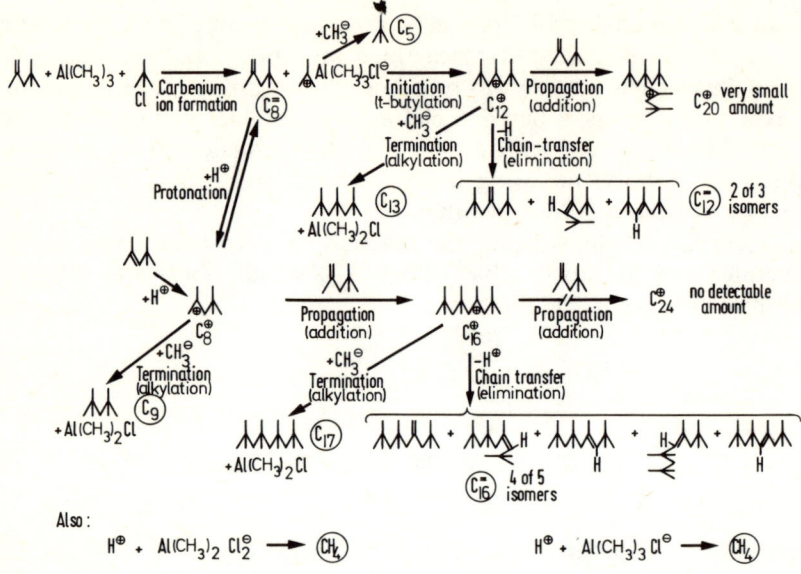

Fig. 1. Kennedy-Gillham model olefin polymerization scheme

These products which arise from initiation i.e., from t-Bu$^{\oplus}$ addition to $C_8^=$ olefin and subsequent reactions of the $C_{12}^{\oplus}$ ion, are termed *first generation products*.

The reaction scheme also considered the olefin distribution arising via proton elimination of the $C_{12}^{\oplus}$ ion: A) 1,1'-dineopentylethylene and B) 2,2',4,6,6'-pentamethyl-3-heptene.

Olefin B, can have either a cis or trans conformation depending on the relative position of the H and CH$_3$ groups.

The proton eliminated from the $C_{12}^{\oplus}$ ion can continue the reaction chain by protonating the $C_8^=$ olefin and thus give rise to a different set of products [Eqs. (13)–(16)]. This process corresponds to chain transfer in polymerization. To distinguish these products resulting from chain transfer from those arising by initiation, the formers are denoted *second generation products.* Thus:

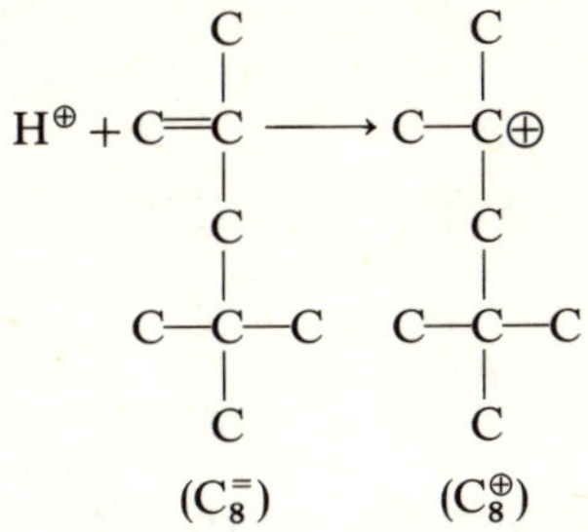

$$H^\oplus + C{=}C \begin{matrix} C \\ | \\ C \\ | \\ C-C-C \\ | \\ C \end{matrix} \longrightarrow \begin{matrix} C \\ | \\ C-C^\oplus \\ | \\ C \\ | \\ C-C-C \\ | \\ C \end{matrix} \qquad (13)$$

$$(C_8^=) \qquad (C_8^\oplus)$$

$$C_8^\oplus \xrightarrow{+C_8^=} C_{16}^\oplus \xrightarrow{-H^\oplus} C_{16}^= + [H^\oplus + Me_3AlCl^\ominus] \qquad (14)$$

$$C_8^\oplus \xrightarrow{Me_3AlCl^\ominus} C_8Me \ (C_9) \qquad (15)$$

$$\xrightarrow{-H^\oplus} C_8^= + [H^\oplus Me_3AlCl^\ominus] \qquad (16)$$

The $C_{16}^\oplus$ ion cannot propagate due to steric hindrance and therefore will yield either $C_{16}^=$ by chain transfer or C_{17} by termination. The $C_{16}^=$ olefin formed by proton elimination from the $C_{16}^\oplus$ ion will be a mixture of five isomers:

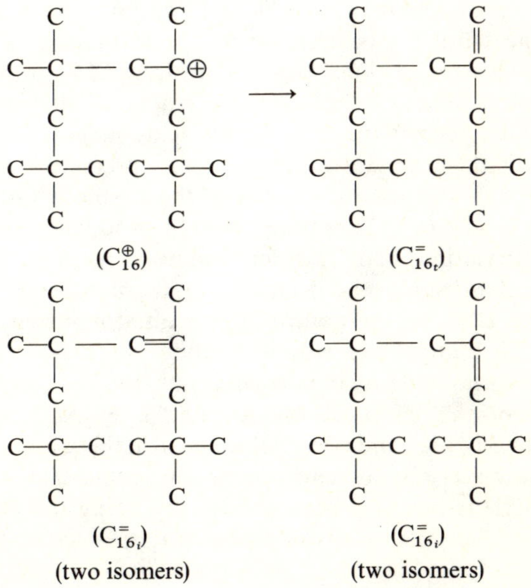

$(C_{16}^\oplus)$ $\quad\longrightarrow\quad$ $(C_{16_t}^=)$

$(C_{16_t}^=)$ (two isomers) $\qquad$ $(C_{16_t}^=)$ (two isomers) where the subscripts t and i signify terminal and internal olefins, respectively. The scheme is completed by the following two reactions:

$$\underset{(t\text{-Bu}^\oplus)}{\overset{\text{C}}{\underset{\text{C}}{\text{C}-\overset{|}{\underset{|}{\text{C}}}\oplus}}} \text{Me}_3\text{AlCl}^\ominus \rightarrow \underset{(\text{C}_5)}{\overset{\text{C}}{\underset{\text{C}}{\text{C}-\overset{|}{\underset{|}{\text{C}}}-\text{C}}}} + \text{Me}_2\text{AlCl} \qquad (17)$$

$$\text{H}^\oplus \text{Me}_3\text{AlCl}^\ominus \rightarrow \text{CH}_4 + \text{Me}_2\text{AlCl} \qquad (18)$$

Equation (17) yielding neopentane and Eq. (18) yielding methane are similar to termination reactions.

Kennedy and Gillham established the validity of this scheme by identifying and quantitatively determining all the products and by internally consistent material balance. In particular, they found that A. Monomer conversion was $\approx 90\%$, B. The reaction was accompanied by methane gas evolution, C. The product mixture consisted of both first and second generation products i.e., C_9, $C_{12}^=$, C_{13}, $C_{16}^=$, C_{17}, and only insignificant amounts of higher fractions. All the compounds were identified by gas chromatography and NMR Spectroscopy, D. Only $<2.0\%$ of the products had more than 17 carbon atoms per molecule. E. Part of the t-BuCl was used up in the formation of neopentane. F. Complicating side reactions such as cracking and hydride ion abstraction were absent. The absence of these reactions alleviates analytical difficulties which occur with other Lewis acids such as $AlCl_3$ or $AlBr_3$.

The excellent correlation between the predicted and obtained products supports the general validity of the Kennedy-Gillham scheme and provides a way to differentiate between products arising by "true" initiation (t-butylation) and "transfer" initiation (protonation).

In the present study the Kennedy-Gillham scheme was used to investigate the effect of the nature of the initiator system, solvent and temperature on competitive reactions under simulated cationic polymerization conditions. More in particular our study concerns a sytematic investigation of the effect of Me_3Al, Et_3Al, Et_2AlCl and Me_2AlCl coinitiators, and t-BuCl and t-BuBr initiators on these reactions in MeCl, MeBr, EtCl and CH_2Cl_2 solvents at different temperatures. Polymerization of isobutylene has also been carried out using the above initiator systems to correlate the results of model reactions with polymerization reactions.

III. Experimental Part

A. Materials

2,4,4-Trimethyl-1-pentene, t-butyl chloride, t-butyl bromide, n-nonane, isobutylene, methyl chloride, ethyl chloride, methyl bromide, methylene chloride and n-hexane were obtained in high purity and further purified by standard methods before use.

The alkylaluminums were obtained from Texas Alkyls Inc. and purified by vacuum distillation in nitrogen atmosphere. B.P. °C/mm Hg: Me_3Al: 60°/68.5; Et_3Al: 97°/100; Et_2AlCl: 125–126°/50 mm; Me_2AlCl: 84°/200. Prior to distillation the dialkylaluminum chlorides were stirred over dry sodium chloride at 80° C for 2 hrs. to remove alkylaluminum dichlorides. The distilled dialkylaluminum chlorides were stored over sodium chloride to prevent the accumulation of the dihalide.

$MeAlCl_2$ (Texas Alkyls Inc.) was obtained as a 25% solution in n-hexane, filtered under nitrogen to remove suspended impurities. The hexane was distilled off and the $MeAlCl_2$ was purified by repeated crystallization from n-pentane at $-50°$. For polymerization reactions $< 1.0\%$ solution in methyl chloride was used.

All experiments were conducted in a stainless steel drybox under nitrogen atmosphere (moisture < 50 ppm). All reactions were carried out in pyrex test tubes with septum caps. The glassware was dried at 140° C for at least 24 h and cooled under nitrogen.

B. Procedure

To a clean dry test tube at the desired temperature, calculated amounts of methyl chloride, 2,4,4-trimethyl-1-pentene, n-nonane and the alkylaluminum compound were added and the test tube was stoppered with a rubber septum cap. The content of the test tube was mixed by shaking and the calculated amount of t-butyl chloride in methyl chloride solvent was rapidly added with a syringe. After ~ 25 min the reaction was terminated by slow addition of prechilled methanol. During termination the septum cap was removed to prevent pressure build-up due to the formation of gaseous products. The tube was removed from the drybox and the solvent was slowly evaporated. To facilitate the separation of the organic products from the aluminum oxide residues a saturated aqueous sodium-potassium tartrate solution was slowly added at 0° C. On adding excess water, the aluminum oxide-tartrate complex can be dissolved to yield well separated aqueous and organic layers. The

Table 1. Product characterization by gas chromatography, molecular weight and proton magnetic resonance spectroscopy[a]

Compound	Symbol	Retention time (sec)	Molecular weight Calc.	Molecular weight Found	Chemical shifts for protons[b]		
2,2′,4,4′-tetramethylpentane	$(C_9)^c$	420	128.2	127—129	$(CH_3)_3$ —CH_2—	1.0s; 1.3s;	18H 2H
2,2′,4,4′-tetramethylhexane	$(C_{10})^{c,d}$	560	142.0	140—142	$(CH_3)_3$ $C(CH_3)_2$ —CH_2— CH_2—CH_3 overlaps with other CH_3 peaks	0.94s; 0.9s; 1.2m;	9H 6H —C—CH_3:
2,2′,4,6,6′-pentamethylheptane	$(C_{12}H)^c$	1080	170.3	169.0—170.6	$(CH_3)_3C$ $C \cdot CH_3$ —CH_2— C—H	0.92s; 1.0s; 1.2m; 1.8m;	18H 3H 4H 1H
1,1′-dineopentylethylene	$(C_{12_t}^=)^c$	1136	168.3	166.4—169.3	$(CH_3)_3C$ $CH_2 \cdot C(CH_3)$ CH_2=C	0.97s; 1.97s; 4.8s;	18H 4H 2H
2,2′,4,6,6′-pentamethyl-3-heptene	$(C_{12_{i'}}^=)^c$	1164	168.3	166.4—169.3	$(CH_3)_3C$ $(CH_3)_3C$—C CH_3C=C —CH_2— C=CH	0.9s 1.1s 1.8d 1.85d 5.1m	9H 9H 3H 2H 1H
2,2′,4,4′,6,6′-hexamethylheptane	(C_{13})	1812	184.4	182.3—187.0	$(CH_3)_3C$ $C(CH_3)_2C$ —CH_2—	1.0s 1.1s 1.36s	18H 6H 4H
2,2′,4,6,6′-pentamethyl-4-ethylheptane	(C_{14})	2280	198.4	195.6—197.6	$(CH_3)_3C$ CH_3—C C—CH_2—C	1.0s 1.06s 1.34m	18H 6H 6H
Mixture of $C_{16}^=$ isomers (3 major peaks)	$(C_{16}^=)^e$	2688, 2820, 2900	224.4	222.9—230.6	CH_2=C C=CH	4.85s 5.1m	2H 1H

Table 1. Product characterization by gas chromatography, molecular weight and proton magnetic resonance spectroscopy[a]

Compound	Symbol	Retention time (sec)	Molecular weight Calc.	Molecular weight Found	Chemical shifts for protons[b]	
2,2′,4,4′,6,6′,8,8′-octamethylnonane	(C_{17})	3240	240.5	240.1—244.2	$(CH_3)_3C$ $C(CH_3)_2C$ $CH_2 \cdot C(CH_3)_3$ $(CH_3)_2C \cdot CH_2$ $\cdot C(CH_3)_2$	1.0s 18H 1.1s 12H 1.34s 4H 1.43s 2H
Higher than C_{17}		3456—4248				

[a] For analysis conditions see Section III C.
[b] s = singlet; d = doublet; m = multiplet.
[c] Correspond with authentic samples.
[d] Prepared according to the reaction $C_8-Cl + Et_3Al \xrightarrow{-78°} C_8-H + C_8-C_2H_5$ (C_{10}).
[e] PMR analysis on olefin mixture for determining isomer ratio only.

products were extracted with n-hexane, dried over anhydrous sodium sulfate, and analyzed by gas chromatography.

For the identification of low boiling fractions, reactions were carried out in two-necked flasks and the low boiling fractions were collected at $-78°$ or below and analyzed by gas chromatography.

Isobutylene polymerizations were carried out by charging isobutylene, methyl chloride, and the alkylaluminum compound in methyl chloride solution and adding the t-butyl halide initiator in methyl chloride rapidly. Polymerizations ensued immediately and were over in 5–10 min. The reactions were terminated after ~ 15 min by the addition of prechilled methanol. The polymers were dried in vacuum at 40° to constant weight and were characterized by number average and viscosity average molecular weights. All reactions were carried out in duplicate.

C. Analyses

A Hewlett-Packard 5750 Research Chromatograph with a flame detector was used. Conditions: High efficiency column with di-n-decylphthalate ($24' \times 1/8''$); carrier gas: helium, injection port: 270° C; detector: 270° C. Column: 25° C till C_9 elution; raised to 70° C in 5 min and held till "C_{12}" elution; raised to 160° C (hold) at 2° C/min.

Quantitative results were obtained using response factors determined with n-nonane internal standard. Peak areas were measured with a K and E planimeter. The low boiling fractions, i.e., methane, ethane, ethylene, isobutane, neopentane and neohexane were identified by authentic samples with a $4' \times 1/8''$ Porapak Q Column. Preparative gas chromatography was carried out with a $6' \times 1/4''$ Apiezon L Column.

Molecular weights were determined using a Chromalytics MC-2 mass chromatograph with SF_6 and CO_2 carrier gases.

NMR spectra of single components or mixtures of isomeric olefins were obtained with a Varian A-60 spectrometer (10% solution in CCl_4; TMS standard).

Analysis of the $C_{12}^=$ and $C_{16}^=$ olefin mixtures by gas chromatography and/or NMR indicated that in both cases the internal isomers were the predominant product. With all the alkylaluminum coinitiators the ratio of internal to terminal olefin was 3:1. Product characterization data are shown in Table 1. Number average molecular weights of polyisobutylenes were determined using a Hewlett-Packard 503 high speed membrane osmometer at 37° C and toluene solvent. Intrinsic viscosities were obtained with a Ubbelohde capillary viscometer at 30° C in cyclohexane solvent. The viscosity average molecular weights were calculated using the relation:

$$\log \bar{M}_v = 5.159 + 1.45 \log[\eta]. \tag{16}$$

IV. Results and Discussion

A1. The Validity of the Kennedy-Gillham Scheme

The present investigation commenced by a reexamination of the scheme proposed by Kennedy and Gillham (9). Thus, t-BuCl initiator was added to a quiescent mixture of 2,4,4,-trimethyl-1-pentene and Me_3Al in methyl chloride at $-78°$ C. The products were quantitatively separated and identified. In contrast to Kennedy and Gillham, who reported high conversions ($\approx 90\%$), only very low ($\sim 0.1\%$) conversions were obtained in spite of repeated attempts in this work. Indeed, conversions remained low ($\approx 1.0\%$) even after quadrupling the t-BuCl

Table 2. Product distribution from the reaction

$$C_8^= + Me_3Al + t\text{-BuCl} \xrightarrow[-78°]{MeCl} \text{products}$$

	Product (moles × 10⁵)			
	Present work			Data from Ref. (9)
	A	B	C	D
Conversion (%)	≤0.1	0.94	≥95.0	88.3
C_1	Tr	Tr	N.D.	~100.0
C_5 [a]	~120.0	~476.0	~23.6	~12.4
C_9	—	—	11.7	5.7
$C_{12}^=$	Tr	2.4	71.9	85.3
C_{13}	Tr	1.0	24.5	30.3
$C_{16}^=$	Tr	<1.0	90.6	86.2
$>C_{16}^=$	Tr	Tr	7.2	10.7

Reaction conditions: $C_8^= = 3.2 \times 10^{-3}$ moles; $Me_3Al = 5 \times 10^{-3}$ moles. Time = 25 min.
A t-BuCl $= 1.2 \times 10^{-3}$ moles; $(Me_3Al/t\text{-BuCl}) \approx 4.0$.
B t-BuCl $= 4.8 \times 10^{-3}$ moles; $(Me_3Al/t\text{-BuCl}) \approx 1.0$.
C Reaction with $C_8^=$ containing 3×10^{-4} moles water and 1.2×10^{-3} moles t-BuCl.

[a] Calculated by differences, i.e., $C_5 = [t\text{-BuCl}] - ([C_{12}^=] + [C_{13}])$.
Tr = Traces   N.D. = Not determined

concentration (see legend under Table 2). Although, the overall yields were low, complete product distribution analyses have been performed. The results of these experiments together with those of the earlier authors are shown in Table 2.

A comparison of the present with the earlier data reveals that while the final conversions are significantly different, the products obtained are identical in both sets of experiments.

Two questions immediately arise from these preliminary experiments: 1. What is (are) the reason(s) for the different conversions and 2. in view of this difference, is the Kennedy-Gillham scheme valid for and can it be used in our intended investigation?

Regarding the first question, Kennedy and Gillham (9) point out that their system contained $\sim 3 \times 10^{-4}$ moles of unaccounted extraneous protons. The source of these protons is obscure. Traces of moisture added with a reagent or moisture condensation while filling the cooled reactor might account for it. While the absolute quantity of these protons is small, it amounts to almost 24% of the total t-BuCl initiator input. This observation indicated the possibility that the differences in conversion levels in the earlier and the present work could have been due to differences in moisture levels.

To verify this, an experiment was carried out with monomer containing $\sim 3 \times 10^{-4}$ moles of water. The results are shown in Table 2. Evidently water significantly increases the conversion levels. While the dry system yields $\sim 0.1\%$ conversion, the "wet" system yields almost complete conversion, similar to the observation of Kennedy and Gillham. The influence of moisture has also been noticed with CH_2Cl_2 solvent. In reactions with rigorously dried (distilled over CaH_2 and stored over molecular sieves) CH_2Cl_2 only $\sim 0.1\%$ conversions were obtained at $-65°$ or below, while the use of undried solvent resulted in high conversions (20%).

The following set of equations provide explanation for the at first glance surprising fact that the "wet" system gives rise to first generation products while neopentane is preferentially formed in the dry system. Water, a strong nucleophile, yields the initiating proton:

$$H_2O + Me_3Al \left(+ \underset{\underset{\underset{C}{|}}{\overset{\overset{C}{|}}{C}-\underset{|}{C}-C}}{C=C} + t\text{-BuCl} \right) \rightleftarrows H^\oplus \, Me_3AlOH^\ominus \qquad (19)$$

which protonates the $C_8^=$ olefin:

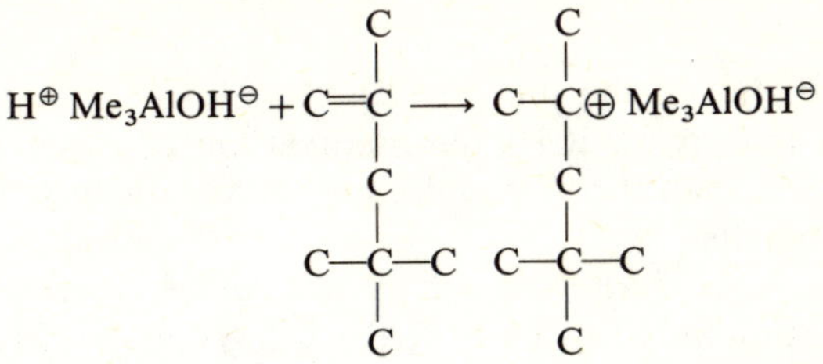

$$H^\oplus \text{ Me}_3\text{AlOH}^\ominus + \underset{\underset{\underset{C}{|}}{\overset{\overset{C}{|}}{C-C-C}}}{\overset{\overset{C}{|}}{C}}=\underset{\underset{\underset{C}{|}}{\overset{\overset{C}{|}}{C-C-C}}}{\overset{\overset{C}{|}}{C}} \longrightarrow \underset{\underset{\underset{}{}}{\overset{\overset{C}{|}}{C-C-C}}}{\overset{\overset{C}{|}}{C}}-\underset{\underset{\underset{}{}}{\overset{\overset{C}{|}}{C}}}{\overset{\overset{C}{|}}{C}}\oplus \text{ Me}_3\text{AlOH}^\ominus \quad (20)$$

The *t*-BuCl in the system presumably acts as a chain transfer agent producing a transitory tertiary chloride and the *t*-butyl cation:

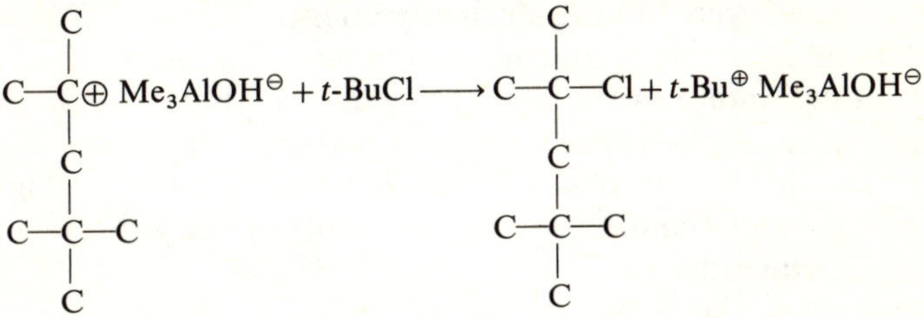

$$\underset{(C_8^\oplus)}{C-C\oplus\atop|\atop C-C-C\atop|\atop C}^{C} \text{ Me}_3\text{AlOH}^\ominus + t\text{-BuCl} \longrightarrow \underset{}{C-C-Cl\atop|\atop C-C-C}^{C} + t\text{-Bu}^\oplus \text{ Me}_3\text{AlOH}^\ominus \quad (21)$$

This reaction can also occur with the "propagating" cation i.e., $C_{16}^\oplus$:

$$C_8^\oplus \text{ Me}_3\text{AlOH}^\ominus + C_8^= \rightarrow C_{16}^\oplus \text{ Me}_3\text{AlOH}^\ominus \quad (22)$$

$$C_{16}^\oplus \text{ Me}_3\text{AlOH}^\ominus + t\text{-BuCl} \rightarrow C_{16}\text{Cl} + t\text{-Bu}^\oplus \text{ Me}_3\text{AlOH}^\ominus \quad (23)$$

The *t*-butyl ion can then *t*-butylate the $C_8^=$ olefin yielding the first generation products:

$$t\text{-Bu}^\oplus \text{ Me}_3\text{AlOH}^\ominus + C_8^= \rightarrow C_{12}^\oplus \text{ Me}_3\text{AlOH}^\ominus \quad (24)$$
$$\phantom{t\text{-Bu}^\oplus \text{ Me}_3\text{AlOH}^\ominus + C_8^= \rightarrow}\hookrightarrow \text{first generation products}$$
$$(C_{12}^=, C_{13})$$

According to this scheme, the $t\text{-Bu}^\oplus$ ion is formed in the presence of $\text{Me}_3\text{AlOH}^\ominus$ counteranion in the "wet" system. In the dry system, however, the t-Bu$^\oplus$ ion is formed in the presence of Me$_3$AlCl$^\ominus$:

$$t\text{-BuCl} + \text{Me}_3\text{Al} \rightarrow t\text{-Bu}^\oplus \text{Me}_3\text{AlCl}^\ominus \tag{25}$$

It is proposed that in the presence of Me$_3$AlOH$^\ominus$, a less nucleophilic counteranion than Me$_3$AlCl$^\ominus$, the methylation of t-Bu$^\oplus$ to form neopentane is slower than the t-butylation of the C$_8^=$ olefin yielding the first generation products.

It remains to be seen whether or not under rigorously dry (high vacuum) condition the t-BuCl + Me$_3$Al + C$_8^=$ system would give t-butylation and subsequent reactions at all or would lead to neopentane exclusively. Or, in the same vein, whether or not under high vacuum conditions the t-BuCl/Me$_3$Al system would polymerize isobutylene (in CH$_3$Cl at $-78°$) at all?

The second question regarding the validity of the Kennedy-Gillham scheme can be answered in the affirmative. The identity of the products obtained in the earlier work and the present investigation, albeit in much lower quantities, indicates the consistency and validity of the Kennedy-Gillham scheme. Further, the internal consistency of this scheme has been substantiated by a large number of experiments carried out under a variety of conditions and to a wide range of conversion levels to be discussed subsequently.

In conclusion, the reexamination of the Kennedy-Gillham scheme revealed that the experiments of the earlier authors have been conducted in the presence of significant amounts of moisture which led to high monomer conversions and small amounts of neopentane formation. The experiments carried out with carefully purified reagents give much lower conversions and large amounts of neopentane. In spite of this difference in conversion levels, the nature of the products under both sets of conditions are identical which in principle substantiates the Kennedy-Gillham scheme as it stands and provides a basis for investigations on the mechanism of cationic olefin polymerizations.

A2. Extension of the Kennedy-Gillham Scheme to Other Alkylaluminum Compounds

The Kennedy-Gillham scheme has originally been worked out for the Me$_3$Al coinitiator/t-BuCl initiator combination. However, this scheme can be readily extended to cover other alkylaluminum compounds. The scheme can directly be applied to the Me$_2$AlCl/t-BuCl system and the product distributions should be the same with Me$_3$Al and Me$_2$AlCl coinitiators. The reason for this identity in products is that both

Cationic Model and Polymerization Reactions of Olefins

counteranions in these systems, $Me_3AlCl^\ominus$ and $Me_2AlCl_2^\ominus$, lead to methylated products; the reactions which do not directly involve the counteranions remain, of course, the same. The *rate* of the reactions will of course be vastly different with Me_3Al and Me_2AlCl, which is explained by the great difference in Lewis acidities of these reagents, but the *nature* of the products will be the same.

The Kennedy-Gillham scheme is also directly applicable to all *t*-BuX initiators. Since the "X" (halogen) becomes attached to the aluminum in the counteranion and since the Al–X bond is relatively strong, the nature of the halogen does not directly affect the identity of the organic products. Again, the rates of the various reactions may be quite different while the nature of the products remains the same.

However, the product distribution of the Kennedy-Gillham scheme must be modified if the organic groups in the alkylaluminum are changed. For example, with Et_3Al or Et_2AlCl and *t*-BuCl ethylation will proceed because the counteranions contain an ethyl group:

$$\begin{array}{c}
\text{C} \\
| \\
\text{C—C}^\oplus \\
| \\
\text{C} \\
(t\text{-Bu}^\oplus)
\end{array} \quad
\begin{array}{c}
Et_3AlCl^\ominus \\
\text{or} \\
Et_2AlCl_2^\ominus
\end{array}
\longrightarrow
\begin{array}{c}
\text{C} \\
| \\
\text{C—C—C—C} \\
| \\
\text{C} \\
(C_6, \text{neohexane})
\end{array} +
\begin{array}{c}
Et_2AlCl \\
\text{or} \\
EtAlCl_2
\end{array} \quad (26)$$

$$\begin{array}{c}
\text{C} \\
| \\
\text{C—C}^\oplus \\
| \\
\text{C} \\
| \\
\text{C—C—C} \\
| \\
\text{C} \\
(C_8^\oplus)
\end{array} \quad
\begin{array}{c}
Et_3AlCl^\ominus \\
\text{or} \\
Et_2AlCl_2^\ominus
\end{array}
\longrightarrow
\begin{array}{c}
\text{C} \\
| \\
\text{C—C—C—C} \\
| \\
\text{C} \\
| \\
\text{C—C—C} \\
| \\
\text{C} \\
(C_{10}, 2,2',4,4'\text{-tetra-methylhexane})
\end{array} +
\begin{array}{c}
Et_2AlCl \\
\text{or} \\
EtAlCl_2
\end{array} \quad (27)$$

$$\begin{array}{c}
\text{C} \quad \text{C} \\
| \quad\; | \\
\text{C—C—C—C}^\oplus \\
| \quad\; | \\
\text{C} \quad \text{C} \\
\quad\;\; | \\
\quad\; \text{C—C—C} \\
\quad\;\; | \\
\quad\;\; \text{C} \\
(C_{12}^\oplus)
\end{array} \quad
\begin{array}{c}
Et_3AlCl^\ominus \\
\text{or} \\
Et_2AlCl_2^\ominus
\end{array}
\longrightarrow
\begin{array}{c}
\text{C} \quad \text{C} \\
| \quad\; | \\
\text{C—C—C—C—C—C} \\
| \quad\; | \\
\text{C} \quad \text{C} \\
\quad\;\; | \\
\quad\; \text{C—C—C} \\
\quad\;\; | \\
\quad\;\; \text{C} \\
(C_{14}, 2,2',4,6,6'\text{-pentamethyl-4-ethylheptane})
\end{array} +
\begin{array}{c}
Et_2AlCl \\
\text{or} \\
EtAlCl_2
\end{array} \quad (28)$$

$$\begin{array}{c} H^{\oplus}\ Et_3AlCl^{\ominus} \\ \text{or} \\ Et_2AlCl_2^{\ominus} \end{array} \longrightarrow \begin{array}{c} C_2H_6 + Et_2AlCl \\ \text{or} \\ EtAlCl_2 \\ (C_2, \text{ethane}) \end{array} \quad (29)$$

$$\begin{array}{c} C_{16}^{\oplus}\ Et_3AlCl^{\ominus} \\ \text{or} \\ Et_2AlCl_2^{\ominus} \end{array} \longrightarrow \begin{array}{c} C_{18} + Et_2AlCl \\ \text{or} \\ EtAlCl_2 \end{array} \quad (30)$$

With iBu_3Al or, in general, with other R_3Al compounds isobutylation and, in general, alkylations can be anticipated. An interesting finding which occurs with alkylaluminum compounds containing a β-hydrogen with respect to aluminum will be discussed later.

Changing the initiator from t-BuX to other suitable organic halides e.g., benzyl halides, acetyl chloride, etc., will also necessitate a modification of the original scheme. The structure of the initiator determines the structure of the head group in the first generation.

In the present investigation the following coinitiator/initiator systems have been studied: Me_3Al/t-BuX, Et_3Al/t-BuX, Et_2AlCl/t-BuX, and Me_2AlCl/t-BuX (X = Cl or Br). In addition, iBu_3Al/t-BuCl and iBu_2AlH/t-BuCl have been employed in special experiments.

A3. Parallelism Between Model and Polymerization Reactions

As discussed in the Model Studies in Cationic Polymerization Section for every reaction in the Kennedy-Gillham scheme a parallel polymerization step exists. By disturbing the basic scheme, quantitative and qualitative effects of changing variables on product distributions can be examined.

Thus, the ratio $C_{12}^{=}/C_{13}$ reflects the relative rate of proton elimination (transfer)/methylation(termination) in the first generation. By changing the coinitiator Me_3Al, Me_2AlCl, the effect of counter anions $Me_3AlCl^{\ominus}$, $Me_2AlCl_2^{\ominus}$ on this ratio can be studied.

Coinitiation efficiency can be (and has been) calculated by the ratio:

$$\frac{[\text{first generation products}]}{[t\text{-BuCl}]}$$

Since steric hindrance prevents propagation ($C_{12}^{\oplus} + C_8^{=} \rightarrow C_{20}^{\oplus}$) the above ratio is a quantitative measure of the coinitiation efficiency ($t\text{-Bu}^{\oplus} + C_8^{=} \rightarrow C_{12}^{\oplus} \rightarrow$ first generation products) and can be studied under various conditions with various initiators.

Propagation is reflected by the quantity of $C_{16}^{=}$ (and higher products) formed. Since the $C_{12}^{\oplus}$ species cannot add $C_8^{=}$ (steric compression), very little or no $C_{20}^{\oplus}$ and subsequent products can arise. However, the $C_8^{\oplus}$ cation formed in the second generation via protonation of $C_8^{=}$, can add one $C_8^{=}$ olefin ($C_8^{\oplus} + C_8^{=} \rightarrow C_{16}^{\oplus}$). Evidently steric compression is not prohibitive for this particular addition. This single olefin addition in the model parallels propagation in polymerization (quasi-propagation). In the present study the propagating abilities of various systems have been analyzed and compared in terms of the $C_{16}^{=}$ and higher fraction formed.

B. Model Studies

Results of a large number of experiments carried out to study the effect of temperature and the nature of the coinitiator in various solvents on conversion and product distributions are summarized in Table 3. The effect of temperature on overall conversions obtained in methyl chloride solvent is plotted separately in Fig. 2. These results will be discussed in terms of the overall coinitiator reactivities and coinitiator efficiencies. Further, the effect of temperature, solvent and the nature of the alkylaluminum coinitiator-alkylhalide initiator system on the fundamental reactions of polymerization will be examined.

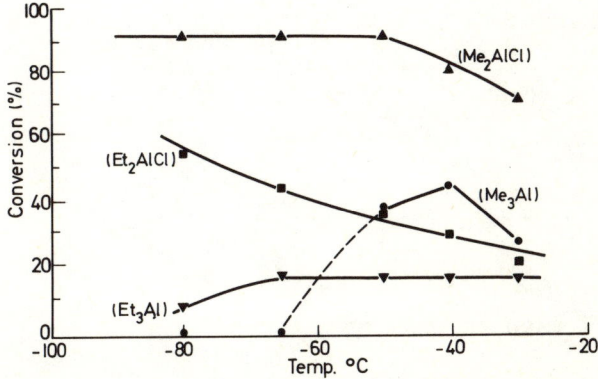

Fig. 2. Effect of temperature on the conversion of $C_8^{=}$ olefin. Initiator = t–BuCl; Solvent = MeCl. Reaction conditions same as in Table 3

Table 3. Effect of temperature on the conversion and product distributions

Temp. °C	−30°				−40°			
Alkylaluminum	Me_3Al	Et_3Al	Et_2AlCl	Me_2AlCl	Me_3Al	Et_3Al	Et_2AlCl	Me_2AlCl
Methyl chloride solvent								
Conv. %	29.2	15.6	20.3	69.8	45.7	17.0	30.9	83.9
Product yield (moles × 10^5)								
C_9	4.2	—	—	6.0	3.1	—	—	5.4
C_{10}	—	4.0	2.2	—	—	3.0	1.8	—
$C_{12}H$	—	31.2	6.0	—	—	42.1	7.8	—
$C_{12}^=$	54.9	1.6	7.9	20.2	76.3	1.8	7.8	22.3
C_{13}	14.0	—	—	0.6	25.4	—	—	0.54
C_{14}	—	2.4	—	—	—	3.5	—	—
$C_{16}^=$	8.8	2.0	24.1	84.8	15.1	1.3	40.2	115.5
$>C_{16}^=$	1.3	Tr	Tr	10.0	1.3	Tr	0.2	4.5
Methylene chloride solvent								
Conv. %	40.4	9.0	26.1	79.4	38.8	11.7	29.5	95.0
Product yield (moles × 10^5)								
C_9	5.9	—	—	5.8	3.0	—	—	8.5
C_{10}	—	1.63	3.0	—	—	1.7	2.5	—
$C_{12}H$	—	19.4	3.6	—	—	25.9	5.64	—
$C_{12}^=$	66.4	2.9	5.1	19.6	61.8	3.2	6.4	23.3
C_{13}	21.9	—	—	0.5	22.7	—	—	0.76
C_{14}	—	1.9	—	—	—	4.4	—	—
$C_{16}^=$	14.6	1.4	36.0	110.0	15.1	1.1	40.0	121.6
$>C_{16}^=$	2.4	Tr	Tr	4.3	2.9	Tr	Tr	9.1
Ethyl chloride solvent								
Conv. %	25.5	11.3	10.4	86.2	31.1	8.8	20.8	95.0
Product yield (moles × 10^5)								
C_9	4.0	—	—	5.5	2.0	—	—	8.2
C_{10}	—	3.2	3.0	—	—	Tr	3.0	—
$C_{12}H$	—	10.5	3.3	—	—	23.0	7.5	—
$C_{12}^=$	55.0	0.5	2.7	19.3	61.0	1.0	5.6	29.8
C_{13}	13.5	—	—	0.69	24.0	—	—	0.9
C_{14}	—	3.0	—	—	—	4.0	—	—
$C_{16}^=$	3.5	1.0	12.1	118.6	5.0	Tr	25.0	128.1
$>C_{16}^=$	1.0	Tr	0.03	6.2	1.0	Tr	0.07	4.2

[a] Reaction conditions: $[C_8^=] = 3.2 \times 10^{-3}$ moles; $[R_3Al] = 4.8 \times 10^{-3}$ moles; $[R_2AlCl] = 3.2 \times 10^{-3}$ moles; $[R_3Al]/[t\text{-BuCl}] = 4.0$; $[R_2AlCl]/[t\text{-BuCl}] = 10.0$. Tr = Traces. Reaction time = 25 min.

obtained with different alkylaluminums/*t*-BuCl initiators in various solvents[a]

−50°				−65°				−80°			
Me₃Al	Et₃Al	Et₂AlCl	Me₂AlCl	Me₃Al	Et₃Al	Et₂AlCl	Me₂AlCl	Me₃Al	Et₃Al	Et₂AlCl	Me₂AlCl
37.3	17.0	34.9	90.0	0.27	16.5	41.0	90.0	0.1	8.9	52.0	95.0
3.0	—	—	4.6	—	—	—	3.7	—	—	—	1.8
—	3.6	2.0	—	—	—	2.2	—	—	—	2.3	—
—	39.9	14.6	—	—	43.8	16.5	—	—	24.6	12.2	—
60.6	2.1	14.4	25.9	1.7	2.8	13.8	26.2	Tr	1.1	11.3	23.7
23.3	—	—	0.86	0.7	—	—	0.86	Tr	—	—	0.6
—	4.2	—	—	—	5.8	—	—	—	2.5	—	—
12.7	1.0	40.0	121.5	Tr	Tr	48.4	118.5	Tr	Tr	76.7	131.5
3.2	Tr	0.22	6.8	Tr	Tr	0.4	9.8	—	Tr	1.1	7.5
43.4	9.8	38.6	95.0	0.6	9.8	51.0	95.0	0.1	0.1	61.6	95.0
2.5	—	—	6.5	—	—	—	9.1	—	—	—	7.3
—	2.3	3.0	—	—	2.8	4.3	—	—	—	3.2	—
—	17.2	8.1	—	—	16.3	11.5	—	—	Tr	10.7	—
68.9	3.8	8.1	28.7	0.7	2.9	10.5	27.3	Tr	Tr	8.4	24.3
30.1	—	—	0.6	0.18	—	—	0.8	Tr	—	—	0.7
—	2.9	—	—	—	2.3	—	—	—	Tr	—	—
15.2	2.4	52.0	130.0	0.6	3.5	68.0	124.3	Tr	Tr	87.0	129.6
3.0	Tr	0.5	8.6	Tr	Tr	0.8	8.9	Tr	Tr	1.0	8.2
5.7	10.4	27.4	95.0	0.15	8.9	50.0	95.0	0.1	2.5	70.1	95.0
—	—	—	6.1	—	—	—	9.7	—	—	—	8.0
—	Tr	2.9	—	—	Tr	4.9	—	—	Tr	2.9	—
—	25.5	10.0	—	—	23.9	12.8	—	—	7.0	19.7	—
12.0	1.5	8.0	27.7	0.34	1.1	9.8	28.0	Tr	0.2	16.2	30.3
4.0	—	—	0.7	0.13	—	—	0.8	Tr	—	—	0.9
—	5.0	—	—	—	3.3	—	—	—	0.8	—	—
1.0	Tr	33.0	126.1	Tr	Tr	65.5	120.7	Tr	Tr	89.5	131.0
Tr	Tr	0.4	8.8	Tr	Tr	0.9	11.8	Tr	Tr	3.0	10.1

B1. Reactivity of Alkylaluminums

Figure 2 shows overall conversions of 2,4,4-trimethyl-1-pentene as a function of temperature obtained with Me_3Al, Et_3Al, Et_2AlCl and Me_2AlCl coinitiators in conjunction with t-BuCl initiator in methyl chloride. Clearly, the activity of the alkylaluminums is greatly affected by temperature, however, the effect is different for each system.

The least reactive coinitiator is Et_3Al. Reactions coinitiated with Et_3Al are least affected by temperature. The increase in temperature from $-80°$ to $-65°$ slightly increases the conversion. Increased amounts of first generation products indicate that this is due to an increased rate of initiation. From $-65°$ to $-30°$ conversions remain fairly constant in any solvent. Et_3Al, the weakest Lewis acid studied, probably gives the most nucleophilic counter anion, and consequently the least separated ion pair which would explain fast ethylation or hydridation (termination) of the t-Bu$^\oplus$.

Me_3Al does not t-butylate 2,4,4-trimethyl-1-pentene below $-65°$, however, produces neopentane indicating very rapid methylation of the tBu$^\oplus$ by the $Me_3AlCl^\ominus$. Evidently at low temperatures methylation (termination) of t-Bu$^\oplus$ is more facile than t-butylation (initiation) of $C_8^=$ olefin. Olefin conversions increase from $-65°$ to $-40°$ but decrease again at $-30°$. The ascending branch at lower temperatures is conceivably due to an increased rate of t-butylation. Indeed, product analysis (Table 3) shows that at temperatures higher than $-65°$, the yields of first generation products ($C_{12}^=$ and C_{13}) increase. At $-65°$ or below mostly neopentane is obtained. The descending branch from $-40°$ to $-30°$ is difficult to explain.

Similar trends have been observed both in methyl chloride and ethyl chloride solvents. In methylene chloride, the overall conversions have not been significantly affected in the range from $-50°$ to $-30°$.

The reactivity of the Et_2AlCl/t-BuCl system decreases monotonically with increasing temperatures. Aside from the maximum range obtained at $-40°$ with Me_3Al, the overall reactivity of Et_2AlCl tends to be higher than that of the trialkylaluminums. Also, higher conversions are obtained with Et_2AlCl than with the trialkylaluminums at, say $-65°$ ($Et_2AlCl = 41\%$, $Et_3Al = 17\%$, $Me_3Al = 0.27\%$), even though the t-BuCl concentration used in conjunction with Et_2AlCl is smaller (Et_2AlCl/t-BuCl $= 10.0$) than that used with the trialkylaluminums (R_3Al/t-BuCl $= 4.0$) (see Table 3). These findings might be due to the stronger Lewis acidity of Et_2AlCl as compared with those of Me_3Al and Et_3Al, however, decreasing conversions with increasing temperatures might also reflect the decreasing polarity of the system.

Cationic Model and Polymerization Reactions of Olefins

Similar to Et_3Al, Et_2AlCl could also terminate by ethylation and/or hydridation especially at higher temperatures. Product analysis (Table 3) tends to support this view. Thus, with increasing temperatures, both the amounts of first generation (reflecting initiation) and second generation ("C_{16}" fraction reflecting propagation) tend to diminish.

Evidently the Me_2AlCl is the most reactive among the four co-initiators investigated and its reactivity is not significantly affected by increasing temperatures. High conversions could be due to the strong Lewis acidity of the Me_2AlCl and indicate slow methylation by the relatively free, stable nucleophile $Me_2AlCl_2^\ominus$. The slight decrase in conversions above $-40°$ can be due to decreased dielectric constant (polarity) of the medium causing the collapse of the ion pairs (termination) at high temperatures. However, even at the highest temperatures the conversions are higher than those obtained with the other alkylaluminum coinitiators.

In sum, dialkylaluminum halides are more reactive than trialkyl-aluminums. The effect of temperature is different for each initiator system and overall conversions are affected by the extents of both initiation and termination.

B2. Coinitiator Efficiency of Alkylaluminums

The following equations summarize the mechanism of initiation and the formation of the first generation products (with Me_3Al):

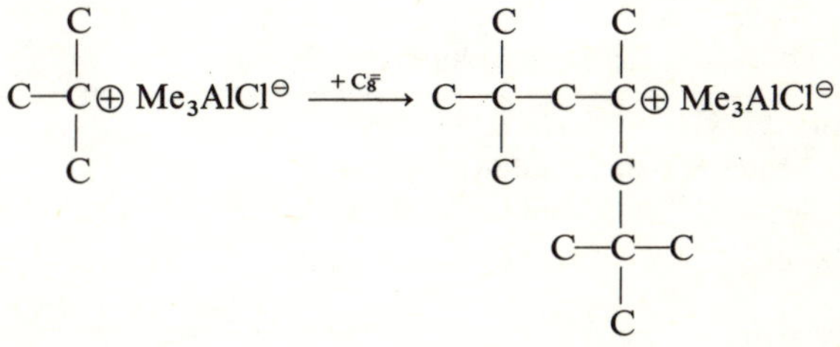

$$\underset{(t\text{-}Bu^\oplus)}{\overset{\underset{|}{C}}{\underset{\underset{|}{C}}{C}}-C^\oplus\ Me_3AlCl^\ominus} \xrightarrow{+C_8^=} \underset{(C_{12}^\oplus)}{C-\underset{\underset{|}{C}}{\overset{\underset{|}{C}}{C}}-C-\underset{\underset{\underset{|}{C}}{C-\underset{\underset{|}{C}}{C}-C}}{\overset{\underset{|}{C}}{C}}\oplus\ Me_3AlCl^\ominus} \qquad (31)$$

$$C_{12}^\oplus \begin{cases} \xrightarrow{+C_8^=} C_{20}^\oplus\ Me_3AlCl^\ominus \text{ (propagation)} & (32) \\ \xrightarrow{-H^\oplus} C_{12}^= + [H^\oplus\ Me_3AlCl^\ominus] \text{ (elimination or transfer)} & (33) \\ \xrightarrow{+Me_3AlCl^\ominus} C_{12}Me + Me_2AlCl \text{ (alkylation or termination)}. & (34) \end{cases}$$

Similar equations can be written for the other alkylaluminum co-initiators.

Since the $C_{12}^{\oplus}$ ion cannot propagate, the efficiency of an initiating system is determined by the ratio $([C_{12}^{=}] + [C_{12}R])/[t\text{-BuCl}]$; where $R = H$, Me or Et. Results are compiled in Table 4. The data show that the general order of coinitiator efficiency is $Me_2AlCl > Me_3Al > Et_2AlCl > Et_3Al$.

The high efficiency of Me_2AlCl is most likely due to its strong Lewis acidity resulting in highly separated ion pairs. Further, $Me_2AlCl_2^{\ominus}$ cannot terminate by hydridation, a most efficient termination mechanism (see later). The low efficiency of Et_3Al is not too surprising since this weak Lewis acid produces relatively tight ion pairs and it can terminate either by ethylation and/or hydridation. The efficiency of Me_3Al and Et_2AlCl tend to fall between the extremes of Me_2AlCl and Et_3Al. The Lewis acidities of Me_3Al and Et_2AlCl also fall between the relatively strong acid Me_2AlCl and weak acid Et_3Al. It could be that the lower efficiency of Et_2AlCl as compared to Me_3Al (at least in the range $-30°$ to $-50°$, where comparative data are available) is due to the fast termination by hydridation with Et_2AlCl, a reaction that cannot take place with Me_3Al.

In general, the coinitiation efficiency tends to decrease at higher temperatures, possibly due to the decreased dielectric constant at higher temperatures. A similar phenomenon has been observed by Kennedy (17) who found higher polystyrene conversions at lower temperatures using the $Et_2AlCl/t\text{-BuCl}$ system in polar solvent, however, only low polystyrene conversions were obtained in nonpolar solvents.

Table 4. Coinitiator efficiency

$$\text{Co.I.E.} = \frac{\text{Moles of "}C_{12}\text{"} + C_{13}(C_{14}) \text{ in product}^{a,b}}{\text{Moles of } t\text{-BuCl initiator added}}$$

Alkylaluminum: Temp. °C	Coinitiator efficiency (%)			
	Me_3Al	Et_3Al	Et_2AlCl	Me_2AlCl
−80	<1.0	29±6	65±7	85±5
−65	<5.0	43±1	55±1	85±5
−50	62±8	35±4	50±6	95±5
−40	83±2	32±8	45±5	75±5
−30	58±1	27±2	38±5	67±7

[a] Reaction conditions as in Table 3.
[b] "C_{12}" includes $C_{12}^{=}$ and $C_{12}H$; C_{13} is obtained with Me_3Al and Me_2AlCl; C_{14} is formed with Et_3Al and Et_2AlCl.

In summary, the coinitiating efficiency of alkylaluminums is determined by their Lewis acidity which in turn controls their terminating ability. Temperature influences the coinitiator efficiency to a significant degree.

B3. Effect of Initiator Systems on Transfer (Elimination) and Termination (Alkylation) A New Termination Mechanism

The relative rates of chain transfer to monomer (briefly transfer), termination and propagation, determine molecular weights and ultimate conversions in most cationic olefin polymerizations. The corresponding model reactions are proton elimination, alkylation by the counteranion and formation of $C_{16}^=$ and higher fractions. Thus, a quantitative analysis of $[C_{12}^=]$, $[C_{13}$ or $C_{14}]$ and $[C_{16}^=]$ could give clues as to the relative rates of these competing reactions. The following equations further illustrate the concept:

Alkylation (termination):

$$C_{12}^\oplus + R_3AlCl^\ominus \rightarrow C_{12}R + R_2AlCl. \tag{35}$$

(R = Me or Et, leading to C_{13} or C_{14})

Elimination (transfer):

$$C_{12}^\oplus R_3AlCl^\ominus \rightarrow C_{12}^= + [H^\oplus R_3AlCl^\ominus]. \tag{36}$$

An important, unexpected observation during the analysis of products obtained with Et_3Al and Et_2AlCl coinitiators led to the postulation of a new general termination mechanism in cationic olefin polymerization. Analysis of the products obtained with the ethyl group-containing coinitiators consistently showed the presence of large amounts of an unexpected compound whose presence was not predicted by the Kennedy-Gillham scheme. The yield of this product was very high ($\approx 95\%$) in Et_3Al coinitiated reactions and had chromatographic retention times close to the $C_{12}^=$ olefins. The product was separated by preparative gas chromatography and analyzed by NMR spectroscopy. The analysis showed the absence of unsaturation, eliminating the possibility of a $C_{12}^=$ isomer. The molecular weight of this compound (170.0) indicated the presence of a C_{12} hydrocarbon ($C_{12}H$). The identity of this compound,

2,2′,4,6,6′-pentamethylheptane, was confirmed by comparing the NMR, IR spectra and gas chromatographic retention times (peak enhancement) with an authentic sample. On the basis of this data, it is postulated that 2,2′,4,6,6′-pentamethylheptane, arises via hydridation of the $C_{12}^{\oplus}$ by the counteranion:

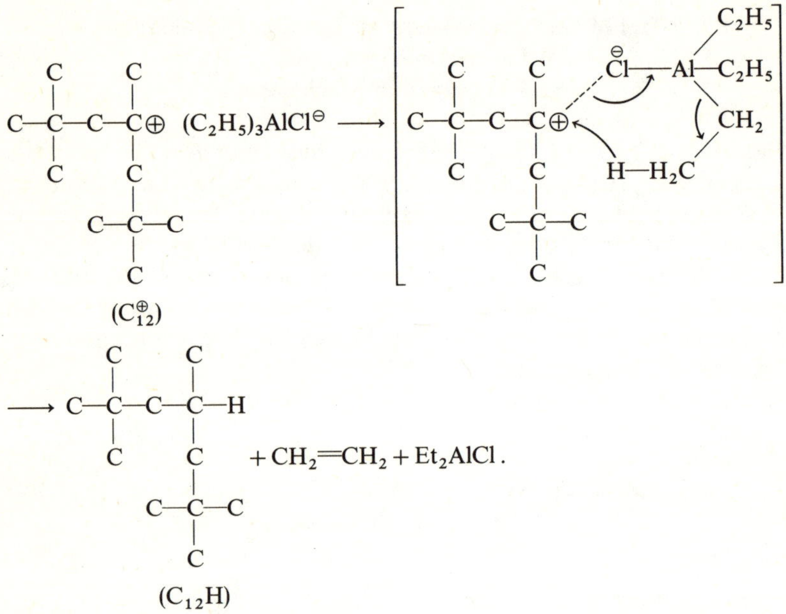

$$\begin{array}{c} \text{C} \quad \text{C} \\ | \quad | \\ \text{C—C—C—C}^{\oplus} \; (C_2H_5)_3AlCl^{\ominus} \longrightarrow \\ | \quad | \\ \text{C} \quad \text{C} \\ | \\ \text{C—C—C} \\ | \\ \text{C} \\ (C_{12}^{\oplus}) \end{array} \quad \left[\begin{array}{c} \text{C} \quad \text{C} \quad \overset{\ominus}{\text{Cl}} \text{—Al—C}_2H_5 \\ | \quad | \quad \quad \quad \backslash \\ \text{C—C—C—C}^{\oplus} \quad \quad CH_2 \\ | \quad | \quad \quad \text{H—H}_2\text{C} \\ \text{C} \quad \text{C} \\ | \\ \text{C—C—C} \\ | \\ \text{C} \end{array} \right]^{\ddagger}$$

$$\longrightarrow \begin{array}{c} \text{C} \quad \text{C} \\ | \quad | \\ \text{C—C—C—C—H} \\ | \quad | \\ \text{C} \quad \text{C} \quad + CH_2{=}CH_2 + Et_2AlCl. \\ \quad \quad | \\ \quad \text{C—C—C} \\ \quad \quad | \\ \quad \quad \text{C} \\ (C_{12}H) \end{array} \quad (37)$$

A search through the literature revealed several instances where a similar mechanism has been postulated (8, 18, 19). It has also been shown that the relative ratios of hydridation and ethylation are affected by the steric requirement of the carbenium ion (18). Highly substituted carbenium ions tend to be terminated by hydridation. Indeed, Priola and coworkers (8) who recently studied the reaction between t-BuCl and a variety of ethylaluminum compounds at $-78°$, found isobutane among the reaction products. These authors postulated termination by hydridation form ethylaluminum compounds at $-78°$ C and summarized their thoughts by the following equation:

$$\begin{array}{c} \text{R} \\ | \\ \text{—C}^{\oplus} \; Et_2AlCl_2^{\ominus} \longrightarrow \text{—CH}_2\text{R} + CH_2{=}CH_2 + EtAlCl_2. \\ | \\ \text{H} \end{array} \quad (38)$$

To substantiate directly the existence of hydridation it was theorized that iBu$_3$Al or iBu$_2$AlH, i.e., compounds having easily accessible and rather basic β-hydrogens, should be even better hydride donors than Et$_3$Al or Et$_2$AlCl and that with these compounds termination will be preferentially by hydridation than by sterically unfavorable isobutylation. Thus, two experiments have been carried out in which t-BuCl was added to quiescent mixtures of 2,4,4-trimethyl-1-pentene and iBu$_3$Al or iBu$_2$AlH in CH$_3$Cl at $-65°$ and after work-up the C$_{12}$ fraction was quantitatively analyzed. According to the data shown in Table 5, hydridation is by far the predominant termination event.

On the basis of this analysis, it is postulated that olefin polymerizations induced by alkylaluminum coinitiators containing a β-hydrogen with respect to aluminum may be preferentially terminated by hydridation by the counter anion.

In the present study, kinetic termination in the presence of Et$_3$Al or Et$_2$AlCl is reflected by the formation of C$_{12}$H (by hydridation) and C$_{14}$ (ethylation). With Me$_3$Al and Me$_2$AlCl only one product, C$_{13}$, due to methylation can form. Elimination/termination ratios are shown in Table 6.

According to the $C_{12}^=/C_{13}$ ratios, elimination is much more important with Me$_2$AlCl than with Me$_3$Al. On the other hand, with the two ethylated coinitiators Et$_2$AlCl and Et$_3$Al, the $[C_{12}^=]/[C_{12}H]+[C_{14}]$ (elimination/termination) is <1.0 indicating preferential termination. It is of interest that the absolute value of this ratio is an order of magnitude higher for Et$_2$AlCl than for Et$_3$Al (1.0 versus 0.07). According to ultimate conversion data, Et$_2$AlCl$_2^\ominus$ is "less terminating" than Et$_3$AlCl$^\ominus$. Similarly, Me$_2$AlCl$_2^\ominus$ is "less terminating" than Me$_3$AlCl$^\ominus$.

It appears that the strongest Lewis acid, i.e., the species producing the least nucleophilic counter anion, gives highest conversions and the highest relative amounts of elimination (and vice versa). The weaker nucleophiles (more stable counter anions) give relatively loose ion pairs, $C_{12}^\oplus//G^\ominus$, and thus longer-lived carbenium ions. Increased conversions (i.e., slower termination) with the stronger Lewis acids could be due to the presence of relatively free carbenium ion-counter ion pairs.

Table 5. Termination by hydride ion with (iBu)$_3$Al and (iBu)$_2$AlH (Initiator: t-BuCl; solvent = MeCl; Temp. $°C = -65°$)

Alkylaluminum	C$_{12}$ distribution		
	C$_{12}$H	C$_{12}^=$	C$_{12}^=/$C$_{12}$H
(iBu)$_3$Al	95.0	5.0	0.05
(iBu)$_2$AlH	85.0	15.0	0.17

Table 6. Effect of alkylaluminum on elimination versus termination[a] of $C_{12}^{\oplus}$ ion (initiator: t-BuCl; solvent: MeCl)

Alkylaluminum	Counteranion	$C_{12}^=/C_{13}$	$C_{12}^=/C_{12}H+C_{14}$
Me$_2$AlCl	Me$_2$AlCl$_2^\ominus$	30±5	—
Et$_2$AlCl	Et$_2$AlCl$_2^\ominus$	—	1±0.2
Me$_3$Al	Me$_3$AlCl$^\ominus$	3±1	—
Et$_3$Al	Et$_3$AlCl$^\ominus$	—	0.07±0.02

[a] Termination includes alkylation and hydridation.

This hypothesis could also account for the observed higher elimination tendency with strong Lewis acids. Thus, if the interaction between the cation and counteranion is weak (weakly nucleophilic counter anion), another nucleophile, the olefin in the system, has a better chance to approach the electrophilic site than with the tighter ion-counteranion pairs (strongly nucleophilic counter anion). However, in the system under investigation, steric hindrance prevents propagation, the ultimate union between the cation and olefin. In this sense, the $C_8^=$ olefin behaves as a nucleophile that is able to abstract irreversibly a proton from the carbocation. Counteranions are also able to abstract a proton, however, in this case a reversible equilibrium exists:

$$C_{12}^{\oplus}//G^{\ominus} \rightleftharpoons C_{12}^= + H^{\oplus}G^{\ominus}$$
$$\xrightarrow{C_8^=} C_{12}^= + C_8^{\oplus}G^{\ominus}.$$

The mobile equilibrium is permanently displaced if the olefin enters the mechanism. Even though the olefin is a weaker nucleophile than the counter anion, the former will ultimately prevail and capture the proton because that process is irreversible. Thus, stronger Lewis acids (i.e., less nucleophilic counter anions) will tend to induce proton elimination (transfer) in preference to alkylation (termination).

To gain more information as to the effect of counter anions on the relative rates of elimination and termination, t-BuBr initiator has been used instead of t-BuCl at selected temperatures in MeCl, MeBr and CH$_2$Cl$_2$ solvents. The use of t-BuBr initiator is expected to yield a counter anion with a bromine instead of a chlorine atom. In particular, the object of these experiments was to determine $C_{12}^=/C_{13}$ and $C_{12}^=/C_{12}H+C_{14}$ ratios in a series of runs and thus to compare the effect of the nature of the halogen in the counter anion on the relative rates of elimination versus termination. Results are shown in Table 7.

With Me_3Al and Et_2AlCl coinitiators, the elimination/termination ratios are 2 to 3 times (range 1.9–3.4) larger in the presence of bromine containing counter anions than those with chlorine containing species. However, no significant differences are observed with Et_3Al and Me_2AlCl coinitiators (cf. Tables 3 and 9).

It could be that with the weakest nucleophile $Me_2AlX_2^\ominus$, the nature of the halogen does not affect ion separation because both $R^\oplus//Me_2AlCl_2^\ominus$ and $R^\oplus//Me_2AlBrCl^\ominus$ are already sufficiently separated to affect the elimination/termination ratio. In contrast, with the strongest nucleophile, $Et_3AlX^\ominus$, the change in halogen from $Et_3AlCl^\ominus$ to $Et_3AlBr^\ominus$ is insufficient to increase the separation of the relatively tight carbenium ion-counter anion pair.

With Me_3Al and Et_2AlCl, coinitiators of intermediate Lewis acidity, the change from chlorine to bromine in the counter anion increases the carbenium ion-counter anion separation and thus gives corresponding increases in the elimination/termination ratio. The solvents examined do not seem to affect this trend.

B4. Effect of Alkylaluminums on Propagation

While steric hindrance prevents the addition of $C_{12}^\oplus$ to $C_8^=$ (propagation), the addition of $C_8^\oplus$ to $C_8^=$ is possible (quasi-propagation) and gives rise to $C_{16}^\oplus$. The $C_{16}^\oplus$ cation in turn mainly eliminates a proton and leads to five $C_{16}^=$ olefin isomers [see Eq. (14)]. In this sense, the yield of $C_{16}^=$ olefins, together with some minor amounts of higher products, can be viewed as a measure of the extent of propagation characteristic of model experiments. More precisely, the yield of $C_{16}^=$ plus higher oligomer fractions, relative to the overall conversion, is a measure of

Table 7. Effect of initiator on elimination versus termination[a] of $C_{12}^\oplus$ ion

Initiating system	Counter anion	Solvents		
		CH_3Cl	CH_3Br	CH_2Cl_2
		$C_{12}^=/C_{13}$ at $-30°$ or $-50°$		
Me_3Al/t-BuCl	$Me_3AlCl^\ominus$	3.5 ($-30°$)	2.5 ($-30°$)	2.3 ($-50°$)
Me_3Al/t-BuBr	$Me_3AlBr^\ominus$	9.0 ($-30°$)	8.0 ($-30°$)	8.5 ($-50°$)
		$C_{12}^=/C_{12}H + C_{14}$ at $-65°$		
Et_2AlCl/t-BuCl	$Et_2AlCl_2^\ominus$	0.8	1.0	0.7
Et_2AlCl/t-BuBr	$Et_2AlClBr^\ominus$	2.1	1.9	1.9

[a] Termination includes alkylation and hydridation.

competition of propagation over termination and/or transfer in the Kennedy-Gillham scheme. The results of typical product analyses along with overall conversions are shown in Table 8.

Reactions coinitiated with Me_2AlCl yield maximum and Et_3Al minimum values. Temperature in the $-30°$ to $-50°$ range have only a minor effect on the overall results. In general terms, the sequence of the rate of "quasi-propagation" is $Me_2AlCl > Et_2AlCl > Me_3Al > Et_3Al$.

This trend as well as those discussed previously in connection with reactivity, coinitiator efficiency, and elimination/termination ratio can be explained by the same principle: The strongest Lewis acid Me_2AlCl which produces the loosest, longest lived carbenium ion pair, produces the largest amount of quasi-propagation, whereas the weakest acid Et_3Al, which gives relatively tight, short-lived carbenium ion pair, produces the least amount of quasi-propagation. The other two species, Et_2AlCl and Me_3Al, fall in between Me_2AlCl and Et_3Al and produce intermediate quantities of $C_{16}^=$ and higher fractions relative to overall conversions.

The available literature on the polymerization of isobutylene using the above coinitiators tends to support this conclusion. Thus, the molecular weights of polyisobutylenes obtained with alkylaluminum coinitiators in conjunction with t-BuCl initiator at any temperature, follow the order $Et_2AlCl > Me_3Al > Et_3Al$ (13). No published data are available for the polymerization of isobutylene with Me_2AlCl. Results with this Lewis acid will be discussed later.

B5. Effect of Solvents

Overall yields and product distributions obtained in experiments using CH_3Cl, CH_2Cl_2, and C_2H_5Cl are shown in Table 3. The overall trends of reactivities obtained with various coinitiators at various

Table 8. Effect of alkylaluminum on propagation (initiator: t-BuCl; solvent: MeCl)

Temp. °C[a]	Et_3Al		Me_3Al		Et_2AlCl		Me_2AlCl	
	Conv. (%)	$\geq C_{16}$ (%)	Conv. (%)	$\geq C_{16}$ (%)	Conv. (%)	$\geq C_{16}$ (%)	Conv. (%)	$\geq C_{16}$ (%)
−30	15.6	8.0	29.2	21.6	20.3	74.2	69.8	84.9
−40	17.0	4.8	45.7	25.1	30.9	82.5	83.9	89.4
−50	17.0	3.7	37.3	26.6	34.9	72.0	>90.0	89.1

[a] Only at these temperatures conversions were sufficiently high with all the coinitiators.
$\geq C_{16}$ (%) indicates percent $C_{16}^=$ and higher fractions.

temperatures are not significantly affected by the nature of the chlorinated polar solvents i.e., CH_3Cl, C_2H_5Cl and CH_2Cl_2. In agreement with these findings, Priola and coworkers (8), who studied the reaction between t-BuCl and ethylaluminum compounds obtained similar results using CH_3Cl and CH_2Cl_2.

B6. Comparison of Alkylaluminum-Alkyl Halide Initiator Systems in Methyl Halide Solvents

Results of experiments using CH_3Cl and CH_3Br solvents with t-BuCl and t-BuBr initiators in conjunction with Me_2AlCl, Et_2AlCl, Me_3Al and Et_3Al are compiled in Table 9. To help surveying the data Fig. 3 shows $C_8^=$ olefin conversions in the form of a histogram.

Reactions coinitiated with Me_2AlCl, be it with t-BuCl or t-BuBr initiators in CH_3Cl or in CH_3Br solvents, always give essentially complete $C_8^=$ conversions (not shown in Fig. 3).

Among the other alkylaluminums the overall sequence of conversions was $Et_2AlCl > Me_3Al > Et_3Al$. The effect of the halogen in the solvent, CH_3Cl or CH_3Br on conversions is also simple: under comparable conditions CH_3Br tends to give lower conversions than CH_3Cl. Changing the halogen from Cl to Br in the initiator, t-BuCl to t-BuBr, gives higher conversions in methyl chloride solvent but lower conversions in methyl bromide.

These observations can be explained along the views discussed earlier. With the strong Lewis acid, Me_2AlCl, changing the counteranion from $Me_2AlCl_2^{\ominus}$ to $Me_2AlClBr^{\ominus}$, i.e., using t-BuCl or t-BuBr, or

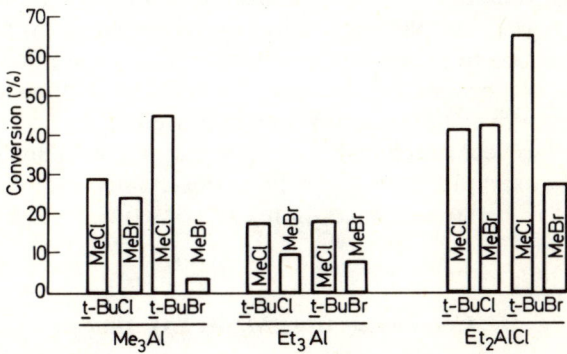

Fig. 3. Histogram showing the combined effect of t–BuX and MeX (X = Cl or Br) on the conversion of $C_8^=$

Table 9. Comparison of alkylaluminum-alkyl

Alkyl–aluminum	Me$_3$Al				Et$_3$Al			
t-BuX	t-BuCl		t-BuBr		t-BuCl		t-BuBr	
CH$_3$X	CH$_3$Cl	CH$_3$Br	CH$_3$Cl	CH$_3$Br	CH$_3$Cl	CH$_3$Br	CH$_3$Cl	CH$_3$Br
Temp. °C	−30°	−30°	−30°	−30°	−50°	−50°	−50°	−50°
Conv. %	29.2	23.9	44.3	2.7	17.0	9.0	17.3	7.5
Product yield (moles × 10^5)								
C$_9$	4.2	0.6	1.2	1.3	—	—	—	—
C$_{10}$	—	—	—	—	3.6	Tr	—	—
C$_{12}$H	—	—	—	—	39.9	22.8	46.8	19.7
C$_{12}^=$	54.9	50.8	82.3	4.2	2.1	1.5	2.9	1.3
C$_{13}$	14.0	20.7	8.7	0.5	—	—	—	—
C$_{14}$	—	—	—	—	4.2	2.48	4.5	2.8
C$_{16}^=$	8.8	1.7	26.2	1.2	1.0	0.9	0.5	0.1
>C$_{16}^=$	1.3	0.3	1.0	0.1	Tr	Tr	Tr	—

[a] Reaction conditions as in Table 3.

changing the solvent from CH$_3$Cl to CH$_3$Br, does not affect the results, presumably because largely dissociated, unencumbered cations are formed under these conditions.

The other alkylaluminum compounds also tend to give overall conversions in line with their Lewis acidities, Et$_2$AlCl > Me$_3$Al > Et$_3$Al. Furthermore, with these relatively weaker Lewis acids, in conjunction with t-BuCl or t-BuBr initiators, changing the solvent from CH$_3$Cl to CH$_3$Br leads to diminished conversions, i.e., poor t-butylation, perhaps because decreased ion pair separation. In other words, in CH$_3$Br preferential alkylation by the counteranion takes place. While no satisfactory explanation can be offered for this observation, it appears that the solvating power of CH$_3$Br is lower than that of CH$_3$Cl.

In CH$_3$Cl solvent the changing of the initiator from t-BuCl to t-BuBr, i.e., going, for example, from R$_3$AlCl$^\ominus$ to the less nucleophilic R$_3$AlBr$^\ominus$, would result in better ion separation and consequently higher conversions.

B7. Conclusions from Model Experiments

1. The simplicity and consistency of the data further substantiates the Kennedy-Gillham scheme.

halide initiators in methyl halide solvents[a]

Et$_2$AlCl				Me$_2$AlCl			
t-BuCl		t-BuBr		t-BuCl		t-BuBr	
CH$_3$Cl	CH$_3$Br	CH$_3$Cl	CH$_3$Br	CH$_3$Cl	CH$_3$Br	CH$_3$Cl	CH$_3$Br
−65°	−65°	−65°	−65°	−65°	−65°	−65°	−65°
41.0	42.2	65.2	27.4	90.0	90.0	95.0	95.0
—	—	—	—	3.7	0.58	2.8	2.3
2.2	2.0	0.3	Tr	—	—	—	—
16.5	9.0	9.2	6.4	—	—	—	—
13.8	9.0	19.8	11.8	26.2	27.9	28.0	27.7
—	—	—	—	0.86	0.6	0.9	0.76
—	—	—	—	—	—	—	—
48.4	56.6	88.4	33.9	118.5	126.7	129.8	127.9
0.43	0.81	1.2	0.7	9.8	2.7	6.5	8.2

2. Product distribution analysis confirms the earlier proposed mechanism of cationic olefin polymerizations using alkylaluminum-alkyl halide systems.

3. Temperature significantly affects the relative rates of reactions and consequently the final yields and product distributions.

4. The nature of the counter anion is determined by the alkylaluminum coinitiator and by the halogen in the t-butyl halide initiator.

5. The activity sequence of the alkylaluminums in terms of overall conversion and coinitiator efficiency is: Me$_2$AlCl > Et$_2$AlCl > Me$_3$Al > Et$_3$Al.

6. The counter anion strongly affects the relative rates of elimination/alkylation (transfer/termination).

7. Alkylaluminums containing a β-hydrogen relative to aluminum preferentially terminate by hydridation, although alkylation might also proceed.

8. The alkylaluminum compound strongly influences termination; the sequence of termination rates based on overall conversions is Et$_3$Al > Me$_3$Al > Et$_2$AlCl > Me$_2$AlCl.

9. The chlorinated solvents, CH$_3$Cl, C$_2$H$_5$Cl and CH$_2$Cl$_2$ do not affect the trends of reactivity of alkylaluminums.

10. The use of methyl bromide solvent in reactions coinitiated with Et$_3$Al, Me$_3$Al and Et$_2$AlCl results in decreased conversions.

11. Reactions coinitiated with the strong Lewis acid, Me_2AlCl, remain unaffected by changing the initiator (t-BuCl or t-BuBr) or the solvent (CH_3Cl or CH_3Br).

These conclusions derived from model experiments lead to the following predictions for isobutylene polymerizations induced by alkylaluminum/t-butyl halide systems.

Among the four alkylaluminums, dialkylaluminum chlorides in conjunction with a suitable alkyl halide will yield highest molecular weight polyisobutylene. The Me_2AlCl/t-BuX system should be the best in this respect. The dialkylaluminum chlorides should have higher coinitiator efficiencies than trialkylaluminums.

It is difficult to predict the molecular weight of polyisobutylene directly from the results of these model experiments. It has been pointed out that the nature of counter anions influences elimination/termination ratios and that by increasing the stability of counter anions the rate of elimination also increases. However, the effect of the nature of counter-anion on propagation (molecular weight) cannot be determined with the model compound. Since the $C_{12}^{\oplus}$ ion cannot add to a $C_8^=$ olefin, it must undergo elimination or alkylation. In contrast to the model olefin, with isobutylene propagation will be favored in the presence of loose ion pairs, and the molecular weights should reflect the relative rates of propagation and transfer.

The high rate of termination that occurs with Et_3Al could be exploited in cationic graft copolymerizations (20). The use of Et_3Al coinitiator could minimize chain transfer due to faster termination, thereby reduce homopolymerization and consequently increase grafting efficiency.

To examine these predictions and to substantiate the Kennedy-Gillham scheme, polymerization of isobutylene has been carried out using Et_3Al, Et_2AlCl and Me_2AlCl coinitiators and t-BuCl or t-BuBr initiators in methyl chloride solvent. The polymerization of isobutylene has also been investigated with methylaluminum dichloride, a Lewis acid which does not require the extra addition of a cationogen for polymerization under "dry box conditions".

C. Polymerization Studies

The results of polymerization studies using isobutylene with Et_3Al, Et_2AlCl and Me_2AlCl coinitiators in conjunction with t-BuCl and t-BuBr initiators, respectively, in methyl chloride solvent are compiled in Table 10, and Figs. 4 and 5. The results obtained with $MeAlCl_2$, a Lewis

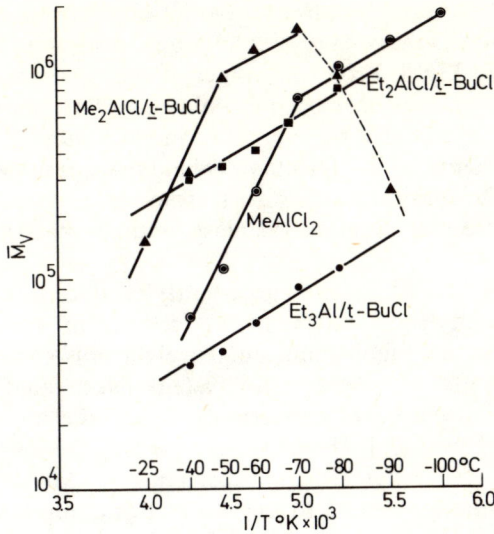

Fig. 4. Effect of temperature on the viscosity average molecular weight of polyisobutylenes, with t–BuCl initiator

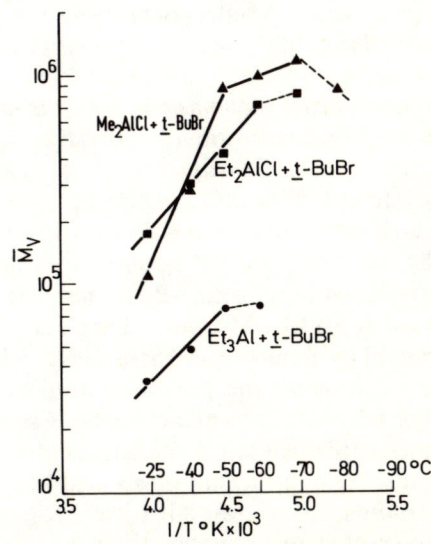

Fig. 5. Effect of temperature on the viscosity average molecular weight of polyisobutylenes, with t–BuBr initiator

acid which does not require the extra addition of a cationogen to induce polymerization of isobutylene under "dry box conditions" are also shown in Table 10 and Fig. 4. From the slopes of the Arrhenius plots (Figs. 4 and 5) the $\Delta E_{\bar{M}_v}$ values have been calculated and are shown in Table 11 along with the available $\Delta E_{\bar{M}_v}$ values in the literature.

In all cases polymer molecular weights increase with decreasing temperature. Exceptions are polymerizations coinitiated with Me_2AlCl, where the molecular weights decrease with decreasing temperature below $-70°$.

The Me_2AlCl/t-BuX system consistently produces highest molecular weight polyisobutylenes. Indeed, in the range from $-50°$ to $-70°$ this system produces the highest molecular weight polyisobutylenes among alkylaluminum-alkyl halide initiator system investigated to date.

The logical extension of these results led to the use of $MeAlCl_2$, a soluble, strong Lewis acid. There are no reports in the scientific literature as to the cationic polymerization behavior of $MeAlCl_2$. The only available information related to the use of this compound is in an old U.S. Patent (21), in which a mixture of equal parts of Me_2AlCl and $MeAlCl_2$ has been employed to polymerize isobutylene at $-78°$. The molecular weights were reported to be in the $10-20 \times 10^3$ range. In the present research $MeAlCl_2$ yielded much higher molecular weight polyisobutylenes. The addition of t-BuCl was unnecessary with this Lewis acid and polymerization started immediately upon $MeAlCl_2$ introduction into the charge. The effect of temperature in the range $-40°$ to $-70°$ on the molecular weight could be expressed by a linear Arrhenius plot, the slope of which gave $\Delta E_{\bar{M}_v} = 7 \pm 1$ kcal/mole. The log molecular weight-reciprocal temperature plot was also linear from $-70°$ to $-100°$, however, the slope was lower corresponding to $\Delta E_{\bar{M}_v} = 2 \pm 0.5$ kcal/mole. Interestingly, this Lewis acid yielded lower molecular weight polyisobytylenes than the Me_2AlCl/t-Bu system between $-40°$ to $-70°$.

Very low conversions are obtained with t-BuBr initiator in conjunction with Et_3Al, Et_2AlCl and Me_2AlCl coinitiators below $-60°$, $-70°$ and $-80°$, respectively, whereas with t-BuCl initiator large amounts of polymer are obtained down to $-80°$ with all these coinitiators. A similar effect has been reported by Kennedy and Sivaram (5) who found bromine is not an effective initiator for the polymerization of isobutylene with Et_2AlCl coinitiators below $-75°$ while chlorine is still effective.

These data can be explained along the lines similar to the discussion of model experiments. With the strong Lewis acid Me_2AlCl, which gives rise to the stable counter anions $Me_2AlCl_2^\ominus$ or $Me_2AlClBr^\ominus$, the molecular weights are expected to be higher than those obtained with the weaker Lewis acids Et_2AlCl or Et_3Al yielding less stable counter anions $Et_2AlCl_2^\ominus$, $Et_3AlCl^\ominus$ or $Et_2AlClBr^\ominus$. In other words, stable counter

Table 10. Polymerization of isobutylene in methyl chloride solvent

Alkylaluminum		Et_3Al[a]		Et_2AlCl[b]			Me_2AlCl[c]		$MeAlCl_2$[d]
Temp. °C	t-BuX	t-BuCl	t-BuBr	t-BuCl	t-BuBr	t-BuCl	t-BuBr	
−25°	Conv. %	—	50±5	—	—	35±5	30±5	—
	$\bar{M}_n \times 10^{-5}$	—	—	—	—	0.76±0.06	0.62±0.01	—
	$\bar{M}_v \times 10^{-5}$	—	0.32±0.02	—	1.7±0.01	1.5±0.2	1.1±0.1	—
−40°	Conv. %	45±5	55±5	45±5	60±5	65±5	75±5	55±3
	$\bar{M}_n \times 10^{-5}$	0.32±0.01	0.35±0.01	0.98±0.01	1.3±0.1	1.4±0.1	1.1±0.1	—
	$\bar{M}_v \times 10^{-5}$	0.39±0.01	0.49±0.01	3.0±0.2	2.9±0.1	3.2±0.2	2.8±0.1	0.65±0.02
−50°	Conv. %	55±5	70±5	65±5	65±5	65±5	80±5	62±2
	$\bar{M}_n \times 10^{-5}$	0.34±0.01	0.45±0.02	1.0±0.05	2.3±0.1	1.8±0.1	2.1±0.1	—
	$\bar{M}_v \times 10^{-5}$	0.44±0.01	0.73±0.01	3.4±0.2	4.2±0.2	9.3±0.6	8.6±0.3	1.1±0.1
−60°	Conv. %	55±5	70±5	75±5	60±5	67±3	70±2	50±2
	$\bar{M}_n \times 10^{-5}$	0.36±0.02	0.47±0.02	1.3±0.1	2.7±0.2	2.3±0.1	2.7±0.2	—
	$\bar{M}_v \times 10^{-5}$	0.61±0.01	0.79±0.01	4.5±0.1	7.1±0.1	12.1±0.3	9.7±0.3	2.5±0.2
−70°	Conv. %	60±5	<10	80±	65±5	84±4	79±3	25±3
	$\bar{M}_n \times 10^{-5}$	0.54±0.02		1.4±0.1	3.1±0.1	3.6±0.2	2.3±0.1	—
	$\bar{M}_v \times 10^{-5}$	0.91±0.01	—	5.9±0.1	8.1±0.2	14.3±0.8	11.1±0.8	7.0±0.2
−80°	Conv. %	40±5	—	70±5	<10	85±5	65±5	18±3
	$\bar{M}_n \times 10^{-5}$	0.62±0.03	—	1.9±0.1	—	1.9±0.2	1.7±0.1	—
	$\bar{M}_v \times 10^{-5}$	1.1±0.1	—	7.8±0.2	—	9.2±0.4	8.4±0.2	10.0±0.5
−90°	Conv. %	—	—	—	—	15±5	<5	15±2
	$\bar{M}_n \times 10^{-5}$	—	—	—	—	0.8±0.1		—
	$\bar{M}_v \times 10^{-5}$	—	—	—	—	2.6±0.2	—	13.6±0.5
−100°	Conv. %	—	—	—	—	—	—	12±3
	$\bar{M}_n \times 10^{-5}$	—	—	—	—	—	—	—
	$\bar{M}_v \times 10^{-5}$	—	—	—	—	—	—	17.5±1.0

Reaction conditions: [Isobutylene] = 3.1 M.
[a] $[Et_3Al] = 6.3 \times 10^{-3}$ M; $[t\text{-BuX}] = 2.5 \times 10^{-3}$ M.
[b] $[Et_2AlCl] = 3.1 \times 10^{-2}$ M; $[t\text{-BuX}] = 1.25 \times 10^{-4}$ M.
[c] $[Me_2AlCl] = 3.1 \times 10^{-2}$ M; $[t\text{-BuX}] = 1.25 \times 10^{-4}$ M.
[d] $[MeAlCl_2] = 1.5 \times 10^{-3}$ M.

Table 11. $\Delta E_{\bar{M}_v}$ for different initiator systems

Initiator system	$\Delta E_{\bar{M}_v}$ kcal/mole	(Range of temp.)	Ref.
Me$_2$AlCl/t-BuCl	7.0 ± 1.0	(−25° to −50°)	This work
	2.0 ± 0.5	(−50° to −70°)	This work
Me$_2$AlCl/t-BuBr	7.0 ± 1.0	(−25° to −50°)	This work
	2.0 ± 0.5	(−50° to −70°)	This work
MeAlCl$_2$	7.0 ± 1.0	(−40° to −70°)	This work
	2.0 ± 0.5	(−70° to −100°)	This work
Et$_2$AlCl/t-BuCl	2.0 ± 0.5	(−40° to −80°)	This work
Et$_2$AlCl/t-BuBr	4.0 ± 0.5	(−25° to −60°)	This work
Et$_3$Al/t-BuCl	2.0 ± 0.5	(−40° to −80°)	This work
Et$_3$Al/t-BuBr	4.0 ± 0.5	(−25° to −50°)	This work
AlCl$_3$	~6.6	(−30° to −100°)	(13)
	~5.6	(−30° to −100°)	(22)
EtAlCl$_2$	~6.6	(−30° to −100°)	(13)
Et$_2$AlCl/t-BuCl	~1.7	(−30° to −100°)	(13)
Et$_2$AlCl/Cl$_2$	~1.9[a]	(−40° to −100°)	(5)
	~7.0[a]	(−30° to −50°)	(23)
Et$_2$AlCl/Br$_2$	~1.9	(−30° to −75°)	(5)
Et$_3$Al/t-BuCl	~1.7	(−50° to −100°)	(13)
Me$_3$Al/t-BuCl	~1.7	(−20° to −78°)	(13)
Me$_3$Al/t-BuCl	~1.9	(−40° to −100°)	(5)
Me$_3$Al/Cl$_2$	~1.9	(−40° to −100°)	(5)
iBu$_3$Al/t-BuCl	~1.7	(−30° to −100°)	(13)

[a] The reason for the difference between the values for the Et$_2$AlCl/Cl$_2$ systems (Refs. 5 and 23) is obscure. It is possible that under the conditions of Ref. 23 the counterion was Et$_2$AlClOH$^\ominus$ i.e., a less nucleophilic species than the expected Et$_2$AlCl$_2^\ominus$.

anions facilitate propagation relative to termination or transfer, thereby yielding high molecular weight polymers.

In regard to the effect of initiator i.e., t-BuCl and t-BuBr, the data indicate that, in conjunction with Et$_3$Al or Et$_2$AlCl, t-BuBr gives higher molecular weights than t-BuCl. Similar results have also been noted by Kennedy and coworkers (5, 14) who obtained higher molecular weight polyisobutylenes with Et$_2$AlCl/Br$_2$ and Me$_3$Al/t-BuBr than with Et$_2$AlCl/Cl$_2$ and Me$_3$Al/t-BuCl initiator systems, respectively. Conceivably, propagation is accelerated in the presence of the larger and consequently more stable bromine-containing counter anions than with the chlorine-containing species.

In the presence of t-BuBr initiator below a certain temperature (see above) coinitiating efficiency precipitously drops to practically nil. This sudden drop is possibly due to very slow initiation at low temperatures. Kennedy and coworkers (7) reported that the rate of neopentane formation via $Me_3Al + t\text{-BuX} \rightarrow t\text{-BuMe} + Me_2AlX$ (X = Cl, Br, I) (and by inference the rate of initiation) depends on the nature of the halogen in t-butyl halide and that the rate follows the order $t\text{-BuCl} > t\text{-BuBr} > t\text{-BuI}$. The differences in the "threshold" temperatures of $-60°$, $-70°$ and $-80°$, i.e., where initiation with $Et_3Al/t\text{-BuBr}$, $Et_2AlCl/t\text{-BuBr}$ and $Me_2AlCl/t\text{-BuBr}$ systems, respectively, suddenly decrease, indicate that coinitiator efficiency (the rate of t-butylation) depend also on the Lewis acidity of the coinitiator.

A decrease in temperature below $-70°$ reduces polyisobutylene molecular weight produced by the $Me_2AlCl/t\text{-BuX}$ (X = Cl or Br) system. It is possible that below $-70°$ the rate of initiation is diminished to such an extent that the unused t-butyl halide starts to function as a chain transfer agent and thus reduces the molecular weights.

The slopes of the Arrhenius plots in Fig. 4 show characteristic differences between the coinitiators Me_2AlCl, $MeAlCl_2$ on the one hand and Et_3Al, Et_2AlCl on the other hand, the former two leading to $\Delta E_{\bar{M}_v} = 7 \pm 1$ kcal/mole and the latter $\Delta E_{\bar{M}_v} = 2 \pm 0.5$ kcal/mole. These values are close to those defined earlier by Kennedy and coworkers (13) for strong Lewis acids, e.g., BF_3, $EtAlCl_2$, $AlCl_3$, etc., ($\Delta E_{\bar{M}_v} = 6.6 \pm 1.0$ kcal/mole) and weaker Lewis acids, e.g., Me_3Al, Et_3Al, Et_2AlCl etc., ($\Delta E_{\bar{M}_v} = 2.0 \pm 0.5$ kcal/mole). The $\Delta E_{\bar{M}_v}$ value of 7 ± 1 kcal/mole for $MeAlCl_2$ is within experimental error of 6.6 ± 1.0 kcal/mole reported earlier for $EtAlCl_2$.

However, the $\Delta E_{\bar{M}_v}$ value of 7 ± 1 kcal/mole obtained with $Me_2AlCl/t\text{-BuCl}$ is higher than expected from the results of $Et_2AlCl/t\text{-BuCl}$. Model experiments indicate that among the four alkylaluminum coinitiators investigated, Me_2AlCl is by far the strongest Lewis acid as reflected in highest conversions and coinitiator efficiencies. These results and the $\Delta E_{\bar{M}_v}$ value obtained in polymerizations indicate that Me_2AlCl (though a dialkylaluminum halide) belongs to the class of strong Lewis acids together with $EtAlCl_2$, BF_3, etc. In contrast to these Lewis acids, carefully purified (i.e., distilled and NaCl treated) Me_2AlCl requires the extra addition of an initiator, e.g., t-BuCl, to induce polymerization even under "dry box conditions". Impure Me_2AlCl (i.e., Me_2AlCl which has not been treated with NaCl) induces the polymerization of isobutylene without the addition of t-BuCl. The possible presence of $MeAlCl_2$ contaminant in Me_2AlCl and traces of moisture may be sufficient to cause polymerization.

Figure 5 shows the Arrhenius plot obtained with t-BuBr initiator in conjunction with Me_2AlCl, Et_2AlCl and Et_3Al coinitiators, respectively. The $\Delta E_{\bar{M}_v}$ value obtained for the $Me_2AlCl/t\text{-BuBr}$ initiator system (7 ± 1 kcal/mole) is similar to that obtained for the $Me_2AlCl/t\text{-BuCl}$ system. This similarity and the similar molecular weights obtained with both systems indicate that with a strong Lewis acid such as Me_2AlCl, changing the initiator from t-BuCl to t-BuBr (that is changing the counter anions from $Me_2AlCl_2^{\ominus}$ to $Me_2AlClBr^{\ominus}$) has no significant effect on the reactivity of the initiator system.

However, $\Delta E_{\overline{M}_v} = 4.0 \pm 0.5$ kcal/mole for polymerizations initiated with t-BuBr in conjunction with Et_2AlCl or Et_3Al. It is of interest that the $\Delta E_{\overline{M}_v}$ values obtained with t-BuBr/Et_2AlCl and t-BuBr/Et_3Al ($\sim 4.0 \pm 0.5$ kcal/mole) fall between the values found to be characteristic for the strong Lewis acids t-BuCl/Me_2AlCl, $EtAlCl_2$, $AlCl_3$, etc. ($\sim 7.0 \pm 1$ kcal/mole) and weak Lewis acid systems t-BuCl/Me_3Al, t-BuCl/Et_3Al, t-BuCl/Et_2AlCl ($\sim 2.0 \pm 0.5$ kcal/mole).

The significance of these $\Delta E_{\overline{M}_v}$ values is that they indicate the quantity $\Delta E_{tr,m}$ for the individual systems. Linear $\log \overline{M}_v$ vs. $1/T$ plots in a temperature range mean that the activation energy difference of the molecular weight determining events remains the same in that range. Since the slope reflects $\Delta E_{tr,m} - \Delta E_p$ and ΔE_p is very low (0–2 kcal/mole), the experimental $\Delta E_{\overline{M}_v}$ values are mainly determined by $\Delta E_{tr,m}$. According to the findings, the highest Arrhenius slope is obtained with the strongest Lewis acids (most stable, least nucleophilic counter anions) which suggests that $\Delta E_{tr,m}$ is highest with these Lewis acids. The most nucleophilic counter anions ($R_3AlCl^{\ominus}$) give the lowest $\Delta E_{tr,m}$ while the t-BuBr/Et_2AlCl and t-BuBr/Et_3Al systems which lead to $Et_2AlClBr^{\ominus}$ and $Et_3AlBr^{\ominus}$, give intermediate $\Delta E_{tr,m}$ values. Evidently the bromine-containing counter anions $Et_2AlClBr^{\ominus}$ and $Et_3AlBr^{\ominus}$ are more stable than the chlorinated ones: $Et_2AlCl_2^{\ominus}$ or $Et_3AlCl^{\ominus}$.

Model experiments have shown that the rate of elimination by the $C_{12}^{\oplus}$ ion increases by changing from $Et_2AlCl_2^{\ominus}$ to $Et_2AlClBr^{\ominus}$. This was explained by the increased stability and consequent better ion pair separation with $R^{\oplus}//Et_2AlClBr^{\ominus}$ over that of $R^{\oplus}//Et_2AlCl_2^{\ominus}$. In the same vein, increased ion pair separation enhances the rate of propagation and gives high molecular weight polyisobutylenes.

The three sets of $\Delta E_{\overline{M}_v}$ values observed in this work indicate that, in addition to the sets of $\Delta E_{\overline{M}_v}$ values defined by Kennedy and coworkers (13), other values are also possible and that these values are determined by the Lewis acidity of the coinitiator and the nature of the initiator. With very strong Lewis acids, no significant differences in $\Delta E_{\overline{M}_v}$ values can be observed possibly because highly dissociated ion pairs are formed under all conditions. With weaker Lewis acids $\Delta E_{\overline{M}_v}$ values are sensitive to subtle changes in the counter anion, for example, changing the halogen from Cl to Br in $Et_3AlX^{\ominus}$ increases the $\Delta E_{\overline{M}_v}$ from ~ 2.0 to ~ 4.0 kcal/mole.

The Arrhenius plots (Figs. 4 and 5) with Me_2AlCl/t-BuX (X = Cl or Br) at or below $-50°$ indicate a change in $\Delta E_{\overline{M}_v}$ from 7 ± 1 to 2.0

± 0.5 kcal/mole, i.e., the $\Delta E_{\bar{M}_v}$ changes from that obtained with strong Lewis acids to that with weaker Lewis acids, such as Et_2AlCl, Me_3Al, etc. The cause for this change in slope is obscure. Evidently, below $-50°$, the molecular weight controlling mechanisms change. It is possible that below this temperature, a different kind of transfer mechanism starts to operate or that termination becomes molecular weight controlling. With $MeAlCl_2$, a similar behavior is noticed, but the change in ΔE_{M_v} occurs below $-70°$.

C1. Comparison Between Model and Polymerization Studies

The effectiveness of Lewis acids Me_2AlCl, Et_2AlCl, and Et_3Al in polymerization of isobutylene follow the trend predicted in model studies, i.e., $Me_2AlCl > Et_2AlCl > Et_3Al$, which supports the conclusions of model experiments. Further, the sequence of coinitiator efficiencies in terms of molecular weights, i.e., $Me_2AlCl > Et_2AlCl > Et_3Al$ is similar to that found in model experiments. The conclusions as to the influence of chlorinated and brominated counteranions (e.g., $Et_2AlCl_2^{\ominus}$ and $Et_2AlClBr^{\ominus}$) found in polymerization follow the expected behavior from the results of model investigations.

However, the model experiments did not give clues for the decreased effectiveness of Me_2AlCl coinitiator observed in polymerizations below $-70°$. The model cannot predict absolute molecular weights in polymerization. Slow initiation in model reaction with Me_2AlCl coinitiator cannot be inferred since the rate of propagation of a $C_8^{\oplus}$ model cation and that by the growing isobutylene cation will be vastly different because of differences in steric requirements. The formation of $C_{16}^{=}$ and higher fraction yields information only about the rate of $C_8^{\oplus}$ addition to a $C_8^{=}$ olefin relative to termination of the $C_8^{\oplus}$. In other words, while the molecular weights obtained are in the millions for isobutylene, mostly dimers are obtained with the $C_8^{=}$ model, independent of the rate of initiation. Thus, the model is insensitive to show the influence of the ratio of the rate of propagation to initiation plus termination.

From the results of model experiments it was predicted that Et_3Al coinitiator in graft copolymerizations will decrease chain transfer, and consequently will maximize grafting efficiency. The basis for this prediction was the finding that termination was fast with $Et_3AlCl^{\ominus}$, leading to termination prior to transfer. Kennedy and Smith (24) proved the validity of this prediction by grafting styrene onto chlorinated ethylene-propylene rubber and obtaining a grafting efficiency of $\approx 90\%$.

The good agreement between the various predictions from the model study and polymerization results supports the conclusions of model experiments and confirms the validity of the mechanism of olefin polymerizations induced by alkylaluminum-alkyl halide systems as proposed by Kennedy and Gillham (9).

V. Summary

Model experiments related to the study of the mechanism of cationic olefin polymerizations have been critically examined. Subsequently the cationation of 2,4,4-trimethyl-1-pentene ($C_8^=$), a nonpolymerizable model olefin for isobutylene, has been investigated by the use of various alkylaluminum compounds i.e., Me_2AlCl, Et_2AlCl, Me_3Al and Et_3Al in conjunction with t-BuCl and t-BuBr initiators using various solvents (CH_3Cl, CH_3Br, C_2H_5Cl, CH_2Cl_2) in the temperature range from $-30°$ to $-80°$ C. Following the cationation of this olefin, the newly-formed carbenium ion undergoes a set of electrophilic reactions, e.g., elimination, alkylation (except addition), and yields well-characterized products, which correspond to a set of the elementary events in cationic polymerization, e.g., initiation, chain transfer, termination (Kennedy-Gillham scheme). In particular, the extended Kennedy-Gillham scheme has been used to study quantitatively the effect of the nature of various alkylaluminum-based initiator systems on reactivity, coinitiator efficiency, elimination/alkylation and quasi-propagation. Comprehensive product analysis indicates that the sequence of reactivities and coinitiator efficiencies is

$$Me_2AlCl > Et_2AlCl > Me_3Al > Et_3Al$$

whereas the sequence of termination rates is exactly the opposite. Temperature significantly affects relative rates and consequently final yields and product distributions, however, the trends cannot be generalized and each system has to be examined individually (Fig. 2). The chlorinated solvents do not affect the results, however, the use of CH_3Br results in decreased reactivity (conversion). Reactions coinitiated with Me_2AlCl are exceptions as they are unaffected by changing the initiator (t-BuCl or t-BuBr) or the solvent (CH_3Cl or CH_3Br). Evidently, the $Me_2AlClX^\ominus$ counter-ion is of very low nucleophilicity and leads to largely unencumbered carbenium ions. An unexpected perturbation in the Kennedy-Gillham scheme and the conclusive identification of an unanticipated hydrocarbon (2,2′,4,6,6′-pentamethylheptane) among the products of the reaction $Et_3Al + t$-BuCl $+ C_8^=$ resulted in the proposition of a new termination mechanism, i.e., termination by hydridation. Termination by hydridation is more important than termination by alkylation if the alkylaluminum coinitiator contains a β-hydrogen in respect to Al. These extensive model experiments led to predictions for polymerization systems that have been subsequently explored by corresponding direct isobutylene polymerizations.

Polymerization of isobutylene was carried out with Et_3Al, Et_2AlCl and Me_2AlCl coinitiators in conjunction with t-BuCl and t-BuBr initiators, respectively, in CH_3Cl in the range from $-25°$ to $-100°$ C. In addition, $MeAlCl_2$, a Lewis acid that does not require the extra addition of a cationogen under "dry box conditions", was also used. In all cases molecular weights increase with decreasing temperatures. The Me_2AlCl/t-BuX produces highest molecular weight polyisobutylenes. The $\log \bar{M}_v$ versus $1/T$ plots were linear for all systems over a considerably wide temperature range. Interestingly, the $\Delta E_{\bar{M}_v}$ values calculated from the slopes of the Arrhenius plots gave either 7.0 ± 1.0 or 2.0 ± 0.5 kcal/mole for alkylaluminum/tBuCl initiating systems. It seems that relatively strong Lewis acids, or, polymerizations with relatively weakly nucleophilic $G^{\ominus}$, give the high $\Delta E_{\bar{M}_v}$ value whereas relatively weak Lewis acids give the low $\Delta E_{\bar{M}_v}$ value. On this basis Me_2AlCl is a relatively strong Lewis acid and belongs to the group with $AlCl_3$ or $EtAlCl_2$. While not completely understood, these facts are related to the energetics of chain transfer.

Acknowledgment is made to the donors of the Petroleum Research Fund, administered by the American Chemical Society, for support of this research.

VI. References

1. Kennedy, J. P.: In: Polymer chemistry of synthetic elastomers, Vol. I, Chapter 5A, p. 291. Kennedy, J. P., Tornqvist, E. (Eds.): Wiley-Interscience, New York 1968
2. Sinn, H., Winter, H., Tirpitz, W. W.: Makromol. Chem. **48**, 59 (1961)
3. Tinyakova, E. I., Zuraleva, T. G., Kurongina, T: N., Kirikova, N. S., Dalgoplask, B. A.: Dokl. Akad. Nauk, USSR **144**, 3, 592 (1962)
4. Saegusa, T., Imai, H., Furukawa, J.: Makromol. Chem. **79**, 207 (1964)
5. Kennedy, J. P., Sivaram, S.: J. Macromol. Sci., – Chem. A **7** (4), 969 (1973)
6. —J. Org. Chem. **35**, 532 (1970)
7. — Desai, N. V., Sivaram, S.: J. Am. Chem. Soc. **95**, 6386 (1973)
8. Priola, A., Cesca, S., Ferraris, G.: Makromol. Chem. **160**, 41 (1972)
9. Kennedy, J. P., Gillham, J. K.: Polymer Preprints **12**, 463 (1971)
10. Whitmore, F. C.: Ind. Eng. Chem. **26**, 94 (1934)
11. Kriz, J., Marek, M.: Makromol. Chem. **163**, 155 (1973)
12. — — Makromol. Chem. **163**, 171 (1973)
13. Kennedy, J. P., Milliman, G. E.: Adv. Chem. Ser. **91**, 287 (1969)
14. — Trivedi, P. D.: University of Akron, unpublished results
15. Alberola, A., Delgaelo, J. A., Fernandez, M. I., Lopez, M. C.: An. Quimica **65**, 517 (1969)
16. Krigbaum, W. R., Flory, P. J.: J. Am. Chem. Soc. **75**, 1775 (1953)
17. Kennedy, J. P.: J. Macromol. Sci.-Chem. A **3**, 5, 861 (1969)
18. Miller, D. B.: J. Org. Chem. **31**, 908 (1966)
19. Alberola, A., Delegaelo, J. A., Fernandez, M. I.: An. Quimica **65**, 495 (1969)
20. Kennedy, J. P.: B. P. 1, 174, 323 (July 5, 1967) to Esso Research and Engineering Co.
21. Kraus, C. A.: U.S.P. 2, 220, 390 (Nov. 12, 1940) to Standard Oil Development Co.
22. Kennedy, J. P., Squires, R. G.: Polymer **6**, 579 (1965)
23. Bacaredda, M., Bruzzone, M., Cesca, J., Di Maina, M., Ferraris, G., Giusti, P., Magagnini, P. L., Priola, A.: Chim. Industria **55**, 2, 109 (1973)
24. Kennedy, J. P., Smith, R. R.: Recent Advances in Polymer Blends, Grafts and Blocks, L. H. Sperling, Ed., Plenum Press, in press. New York 1974

Received September 24, 1973

Photoinitiation of Vinyl Polymerization by Aromatic Carbonyl Compounds

J. HUTCHISON and A. LEDWITH

Donnan Laboratories, University of Liverpool, Liverpool L 69 3 BX, England

Table of Contents

I. General Aspects of Light Absorption Processes 50
   A. Excited States and Possibilities for Photoinitiation 50
   B. The Nature and Reactivity of Excited States of Aromatic
      Carbonyl Compounds 52

II. Photoinduced Fragmentation Reactions of Aromatic
    Carbonyl Compounds 56
   A. Benzoin Derivatives 56
   B. Sulphur- and Halogen-Containing Aromatic Carbonyl Compounds . 62

III. Photoinduced Hydrogen Abstraction by Aromatic Carbonyl Compounds 65
   A. General Mechanism 65
   B. Photoinitiation by Benzophenone Derivatives 67
   C. Photoinitiation by Benzil 74
   D. Photoinitiation by Quinones 75

IV. Photoinduced Electron Transfer Reactions of Aromatic
    Carbonyl Compounds 78
   A. Reactions Involving N-Vinyl Carbazole 78
   B. Initiation of Free Radical Polymerization 80

V. References 84

I. General Aspects of Light Absorption Processes

A. Excited States and Possibilities for Photoinitiation

Molecules have minimum electronic energy when in the ground state which, for the vast majority of molecules is a singlet state, symbolised S_0. Promotion of an electron from the highest occupied molecular orbital in the ground state to the lowest vacant molecular orbital is the transition of lowest energy which results in a change in electronic configuration. The state produced by this electronic transition may be either singlet or triplet, and will be the first excited state, symbolised S_1 or T_1 respectively. Singlet states are invariably of higher energy than the corresponding triplet state because of greater electronic repulsion in the former. Excited states of greater energy than S_1 and T_1 (e.g. S_2, S_3, ..., T_2, T_3, ...) may be formed by similar excitation processes, but internal conversion of such states to the corresponding first excited states is normally much more rapid in solution than any normal photochemical process and, in consequence, their existence will be largely ignored for purposes of this survey. Since the majority of molecules exist in ground singlet states, it follows, from the rules governing spin conservation, that direct excitation of such molecules must produce mainly singlet excited states. Once formed, the (lowest) excited singlet state may lose its excitation energy in one of four main processes:

(i) Radiationless conversion back to ground singlet state.

(ii) Radiative conversion back to ground singlet state (fluorescence).

(iii) Quenching of the excited singlet state by interaction with other constituents of the system.

(iv) Conversion to the corresponding triplet excited state (intersystem crossing).

Generation of excited triplet states is normally achieved as a result of (iv) above and, once formed, they may decay by processes analogous to (i)—(iv) with the obvious distinction that radiative decay of triplet states is termed phosphorescence, and that radiationless transition of excited triplet states, back to ground singlet states, involves intersystem crossing. Whilst there are very many mechanisms whereby so called quenching of excited states may occur (*1, 2*), and a full discussion is outside the scope of this article, a large part of the review will be concerned with a particular form of quenching in which appropriate excited states are deactivated by reactions with added substrate or solvent. Radiative lifetimes (τ) of excited singlet and triplet states differ markedly, the latter being larger by many orders of magnitude, and consequently the majority of useful photochemical reactions occur from triplet excited states.

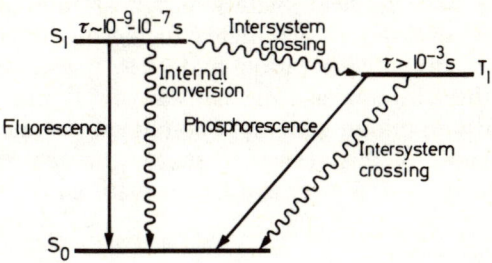

In principle, there are at least five distinct mechanisms by which a photo excited molecule (A*) may initiate polymerization:

a) Direct addition of A* to monomer (M) producing a biradical or dipolar species e.g.

$$A^* + M \rightarrow \cdot A-M \cdot \quad \text{or} \quad {}^+A-M^- \quad \text{or} \quad {}^-A-M^+$$

b) Energy transfer from (mainly) triplet excited molecules to monomer to produce triplet excited monomer e.g.

$$A(T_1) + M(S_0) \rightarrow A(S_0) + M(T_1)$$

c) Homolytic fragmentation of the photo-excited molecule e.g.

$$A-B^* \rightarrow A\cdot + B\cdot$$

d) Hydrogen abstraction by A* from monomer, solvent etc. so as to produce two radicals e.g.

$$A^* + RH \rightarrow \dot{A}H + R\cdot$$

e) Electron transfer between photoexcited molecule and monomer, solvent, etc. so as to produce a pair of ion radicals e.g.

$$A^* + M \rightarrow M^{\pm}, A^{\mp} \quad \text{or} \quad M^{\mp}, A^{\pm}$$

Mechanisms a) and b) yield mainly biradical intermediates which, because of facile cyclisation, are highly inefficient initiating and propagating species in polymerization and these processes will not be considered further. In contrast, mechanisms c), d), and e) may all be used conveniently to initiate (mainly) free radical polymerization and it will be the purpose of this review to survey the general applicability of these processes in so far as they involve aromatic carbonyl compounds.

B. The Nature and Reactivity of Excited States of Aromatic Carbonyl Compounds

Molecules containing heteroatoms have electrons in orbitals, associated with the heteroatom, which are not involved in the bonding system of the molecule. Carbonyl compounds provide good examples since there are two electrons in each of the non-bonding n orbitals of the oxygen atom. Absorption of radiation can lead to the promotion of one of these electrons into either a σ^* or a π^* orbital i.e. $n \rightarrow \sigma^*$ or $n \rightarrow \pi^*$ transitions.

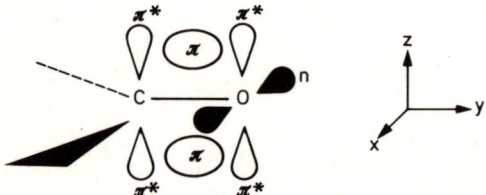

Orbitals associated with a C=O group. The n orbital (represented as a $2p$ orbital) points along the x axis while the π and π^* orbitals lie in the yz plane

For most molecules energies of the various bonding and antibonding orbitals increase in the order

$$\sigma < \pi < n < \pi^* < \sigma^*$$

and this means that the most readily observable electronic transitions are $n \to \pi^*$ and $\pi \to \pi^*$ in nature. Usually $n \to \pi^*$ transitions are lower in energy than corresponding $\pi \to \pi^*$ transitions but the positions may be reversed when the molecule has a high degree of conjugation, in which case the π orbital has a higher energy than the n orbital. This point is of considerable significance when considering photo-reactions of aromatic carbonyl compounds as will become evident from the following discussion. Quite apart from relative energy levels, there are very important differences in the electronic structures and chemical reactivities of (n, π^*) and (π, π^*) excited states. Inspection of the diagrammatic representation of the orbitals of a carbonyl group, shown above, indicates that promotion of an electron from an n orbital to a π^*-antibonding orbital has the effect of *removing* electron density from the oxygen atom with the result that (n, π^*) excited states (especially triplet states) display reactivity similar to that of alkoxy radicals. In particular, hydrogen abstraction reactions — well characterised for alkoxy radicals — are frequently encountered for (n, π^*) excited states. On the other hand, promotion of a π-bonding electron to a π^*-antibonding orbital has the effect of *increasing* electron density at the oxygen atom with a consequent increase in the polar nature of the $>C=O$ group. The differing electron distributions of (n, π^*) and (π, π^*) excited states are manifested in the effects of solvent polarities on the energies of the transitions. Increasing solvent polarity causes a decrease in the energy of $\pi \to \pi^*$ transitions (red shift) largely because polar solvents will reduce the energy of the (more polar) excited state by increased solvation. Corresponding $n \to \pi^*$ transitions are shifted to shorter wavelengths (higher energies) by increasing solvent polarity, especially so by hydrogen bonding solvents. This so-called blue shift results from the decrease in energy of the n orbital consequent on hydrogen bonding, or other dipolar interactions. It should be noted that some aromatic carbonyl compounds, e.g. acetophenone (3), have comparatively small energy separations between highest energy n and π-bonding orbitals and hence may give rise to lowest lying (n, π^*) or (π, π^*) excited states depending on the particular solvation atmosphere.

A further distinction between (n, π^*) and (π, π^*) excited states results from the selection rules for electronic transitions which, as can be seen in the orbital diagram above, classify the former as partially forbidden — essentially because of the poor spatial overlap between n and π^* orbitals, and the latter as allowed. In consequence molar extinction coefficients for typical $n \to \pi^*$ transitions lie in the range $10 - 10^2 \, M^{-1} \, cm^{-1}$ whereas those for $\pi \to \pi^*$ transitions range from $10^3 - 10^5 \, M^{-1} \, cm^{-1}$, a matter of considerable practical significance in applications of photoinitiated polymerizations.

Undoubtedly the most important distinction between (n, π^*) and (π, π^*) excited states of aromatic carbonyl compounds is the much greater efficiency of intersystem crossing (i.e. $S_1 \to T_1$ conversion) in the former cases. This follows because, as already noted, triplet states are the longest lived and most useful intermediates in organic photochemistry. Two factors contribute to the high probability of intersystem crossing between (n, π^*) states as compared to that between (π, π^*) states:

1. The initial $n \to \pi^*$ excitation is partly forbidden and hence the reverse transition will also be partly forbidden. Thus the lifetimes of $S_1(n, \pi^*)$ states tend to be greater than $S_1(\pi, \pi^*)$ states and the probability of conversion to the T_1 state correspondingly higher for $S_1(n, \pi^*)$ states.

2. Energy separations between S_1 and T_1 states of the (n, π^*) type are normally quite small (1500—5000 cm^{-1}) whereas those for the corresponding (π, π^*) states are usually much larger (10000 to 15000 cm^{-1}).

For aromatic carbonyl compounds having lowest energy $n \to \pi^*$ transitions, intersystem crossing is essentially quantitative.

Of the readily available aromatic carbonyl compounds, benzophenone derivatives generally give lowest energy $n \to \pi^*$ transitions, with complete conversion $S_1 \to T_1$, and find extensive use in organic photochemistry in consequence. Carbonyl triplet states of the (n, π^*) nature readily abstract hydrogen atoms from reactive substrates such as alcohols and ethers. These reaction occur with very high quantum efficiencies and are discussed in more detail in Section III. However it is useful to note here that recent theoretical studies (4) confirm the biradical nature of (n, π^*) triplet states and indicate a transition state for hydrogen abstraction involving fractional bond orders e.g.

$$\left[\triangleright\!\!\!C =\!=\!=\!= O - - - H - - - R \right]^{\ddagger}$$

Two types of aromatic carbonyl compound give rise to lowest energy excited states not having (n, π^*) character. Amino substituted benzophenones, especially, *para*-substituted compounds, form excited states which are neither (n, π^*) nor (π, π^*) in character, but which are better described as charge transfer states. A typical case is provided by p-aminobenzophenone which yields an excited state structure as shown:

Whether the charge transfer state is lower in energy than the corresponding (n, π^*) triplet state depends largely on the solvent polarity, and since such compounds have not yet found practical use in polymerization processes they will not be considered further. Whenever the degree of conjugation in aromatic carbonyl compounds is more extensive than, for example, the Ph—$\overset{|}{C}$=O group, the lowest lying triplet excited states are frequently (π, π^*) in character and such compounds do not readily undergo the classical photoinduced hydrogen abstraction with alcohols and ethers. Examples of this type of compound include fluorenone, xanthone, p-phenylbenzophenone, and α- and β-naphthyl carbonyl compounds. Relative energies of singlet and triplet excited states of benzophenone and 2-acetonaphthone are indicated below and it will be seen that whilst the lowest energy singlet excited states are (n, π^*), for both compounds, the lowest energy triplet states are different in character.

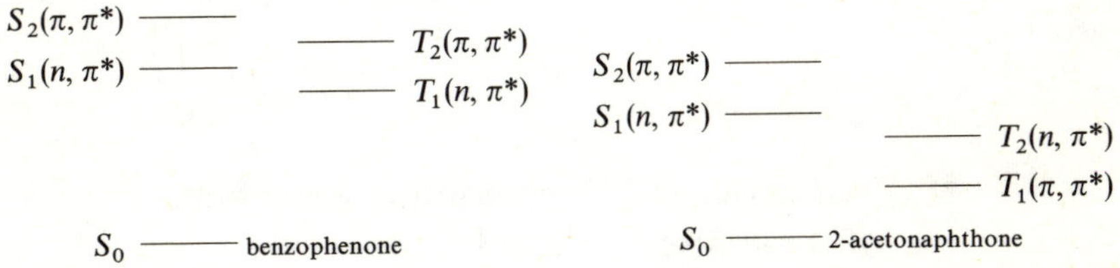

$S_2(\pi, \pi^*)$ ———
$S_1(n, \pi^*)$ ———   ——— $T_2(\pi, \pi^*)$ $S_2(\pi, \pi^*)$ ———
                          ——— $T_1(n, \pi^*)$ $S_1(n, \pi^*)$ ———
                                                                                  ——— $T_2(n, \pi^*)$
                                                                                  ——— $T_1(\pi, \pi^*)$

S_0 ——— benzophenone S_0 ——— 2-acetonaphthone

Whereas aromatic carbonyl compounds having lowest lying (π, π^*) triplet excited states are not photoreduced by hydrogen donors such as alcohols and ethers, they are readily photoreduced by electron donors such as amines and sulphur compounds (5). The electron transfer processes may be generalised as follows:

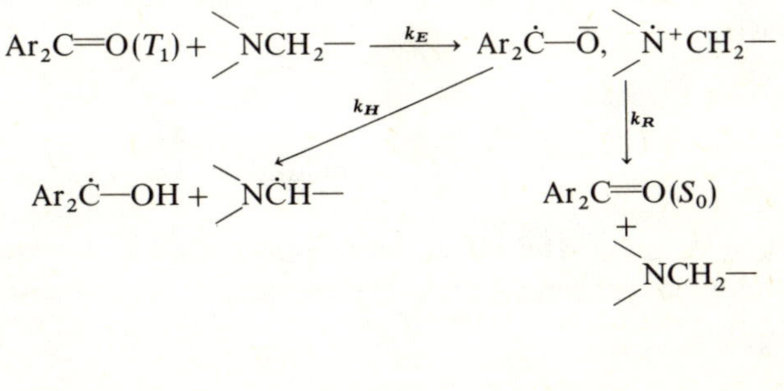

$$Ar_2C\!=\!O(T_1) + {>}\!NCH_2\!- \xrightarrow{k_E} Ar_2\dot{C}\!-\!\bar{O}, \dot{N}^+CH_2\!-$$

$$\swarrow k_H \qquad\qquad \downarrow k_R$$

$$Ar_2\dot{C}\!-\!OH + {>}\!N\dot{C}H\!- \qquad Ar_2C\!=\!O(S_0)$$
$$+$$
$${>}\!NCH_2\!-$$

Although the pair of ion radicals are depicted as arising directly from encounter between triplet excited carbonyl compound and the (amine) electron donor, it is highly likely that individual systems show variations due to intermediate formation of exciplexes such as $[Ar_2C=O]_{T_1} \cdots$ Amine. The reader is referred to a recent review by Ottolenghi (6) for a stimulating account of the role of exciplexes in photoinduced electron transfer reactions. However, the simplified scheme indicated serves to illustrate that formation of a potentially reactive pair of free radicals by proton transfer depends on the relative values of k_E, k_H and k_R. Solvent polarity will have an obvious effect on these processes (and the intermediate formation of exciplexes), and the reactions are favoured by increasing electron availability in the amine. If α-hydrogen is not present in the amine structure, k_H reduces to zero and there will be excited state quenching only (represented by k_R). In any case the competition between k_H and k_R will normally mean that quantum yields for photoreduction by this mechanism are less than the theoretical maximum of 2, achieved in photoreduction of $n-\pi^*$ triplet excited carbonyl compounds with alcohols. A comprehensive survey of photophysical and photochemical effects in aromatic ketone-amine systems has been given by Cohen, Parola and Parsons (5).

II. Photoinduced Fragmentation Reactions of Aromatic Carbonyl Compounds

A. Benzoin Derivatives

Among the aromatic carbonyl compounds which undergo fragmentation when irradiated with u.v. light, benzoin and benzoin alkyl ethers have been the most widely used as photoinitiators for vinyl polymerization (7—15). Industrial applications of these compounds include the manufacture of printing plates (15) (relief and dry offset) and the formulation of printing inks (14), and surface coatings (13) the curing of which involves the light-induced copolymerization of unsaturated polyesters with styrene.

1. Mode of Fragmentation

In the first detailed study of benzoin photoinitiated polymerization of vinyl monomers (styrene, butyl acrylate, methyl acrylate and methyl methacrylate) Melville (16) obtained kinetic evidence for photochemical production of radicals from benzoin. This was assumed to proceed by a Norrish Type I cleavage.

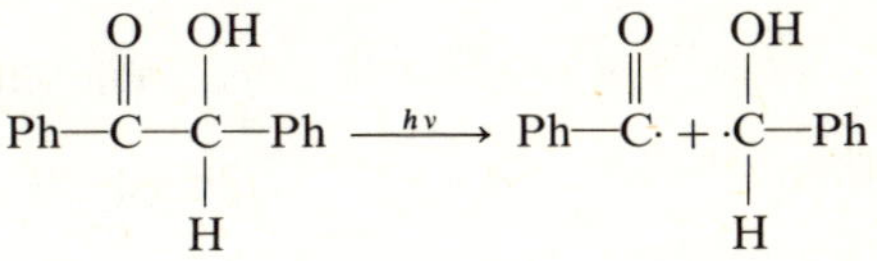

$$\text{Ph-}\underset{\text{H}}{\overset{\overset{\text{O}}{\|}}{\text{C}}}-\underset{\text{H}}{\overset{\text{OH}}{|}}-\text{Ph} \xrightarrow{h\nu} \text{Ph-}\overset{\overset{\text{O}}{\|}}{\text{C}}\cdot + \cdot\underset{\text{H}}{\overset{\text{OH}}{|}}-\text{Ph}$$

Recent CIDNP experiments (17) support this view since it was observed that the radical pair formed on irradiation of benzoin was identical to that formed on irradiation of benzaldehyde.

Further evidence for Type I cleavage has been obtained by the use of diamagnetic radical scavengers as spin traps for radicals produced on photolysis of benzoin and benzoin methyl ether (18). In both cases $Ph\dot{C}O$ and $Ph\dot{C}HOR$ (R = H, Me) radicals were trapped and characterised from the e.s.r. spectra of the stable nitroxide radicals formed e.g.

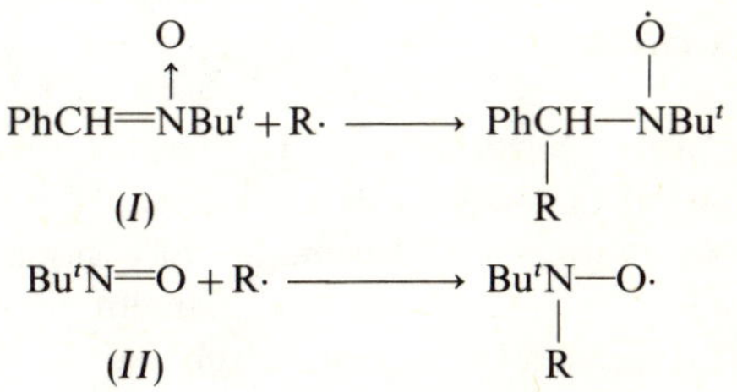

$$\text{PhCH}\overset{\overset{\text{O}}{\uparrow}}{=}\text{NBu}^t + \text{R}\cdot \longrightarrow \text{PhCH}-\underset{\text{R}}{\overset{\overset{\dot{\text{O}}}{|}}{\text{NBu}^t}}$$

(I)

$$\text{Bu}^t\text{N}=\text{O} + \text{R}\cdot \longrightarrow \text{Bu}^t\underset{\text{R}}{\text{N}}-\text{O}\cdot$$

(II)

(Note that these reactions are somewhat analogous to addition of the radicals to vinyl monomers.) Similar results were obtained for reactions carried out in benzene and in methanol, the latter showing that fragmentation of photoexcited benzoin and benzoin methyl ether occurs much more rapidly than hydrogen abstraction from methanol.

Heine (7, 19) isolated the products formed on photolysis of benzoin alkyl ethers (methyl, ethyl, isopropyl) in benzene, and these also were indicative of the intermediacy of benzoyl and alkoxybenzyl radicals:

$$\text{Ph}-\underset{\underset{H}{|}}{\overset{\overset{O}{\|}}{C}}-\underset{}{\overset{\overset{OR}{|}}{C}}-\text{Ph} \xrightarrow{h\nu} \text{Ph}-\overset{\overset{O}{\|}}{C}\cdot + \cdot\underset{\underset{H}{|}}{\overset{\overset{OR}{|}}{C}}-\text{Ph}$$

$$\text{PhCHO} + \text{Ph}-\underset{\overset{\|}{O}}{\overset{\overset{O}{\|}}{C}}-\underset{\overset{\|}{O}}{\overset{\overset{}{}}{C}}-\text{Ph} \qquad \text{Ph}-\underset{\underset{H}{|}}{\overset{\overset{OR}{|}}{C}}-\underset{\underset{H}{|}}{\overset{\overset{OR}{|}}{C}}-\text{Ph}$$

2. Relative Efficiencies

The efficiencies of benzoin derivatives in promoting the light induced hardening of styrene-containing unsaturated polyester resins have been found to vary considerably (7), benzoin alkyl ethers and α-alkylated benzoins being more reactive than benzoin itself. However, as photoinitiators for the polymerization of bulk methyl methacrylate, benzoin and benzoin methyl ether exhibit comparable efficiencies (20), and this is also indicated for polymerisation of methyl acrylate in *tert*-butanol (21). The relative photoinitiating efficiencies of benzoin derivatives may be influenced by several factors, some of which will depend on the nature of the monomer system and the environment in which the polymerizations are carried out.

a) Light Absorption Characteristics. In the bulk methyl methacrylate study (20), the amount of ultra-violet radiation absorbed by the photoinitiator solutions was measured by means of a potassium ferrioxalate actinometer (22). It was found that the benzoin methyl ether system absorbed approximately twice as much light as that absorbed by the same concentrations of benzoin (in the range $O - 5 \times 10^{-2}$ M) and it was concluded that this more than adequately accounted for the generally higher rates of polymerization observed with benzoin methyl ether. That benzoin methyl ether would absorb more of the incident light than benzoin was predictable from the ultraviolet absorption spectra (Fig. 1), since the most intense band emitted by the lamp used in the experiments (a 250 W medium-pressure mercury discharge lamp) was centred around 366 nm. The dissimilarity between the spectra in Fig. 1 may be ascribed largely to the presence of strong intramolecular hydrogen bonding in the benzoin molecule which is not possible in the case of benzoin methyl

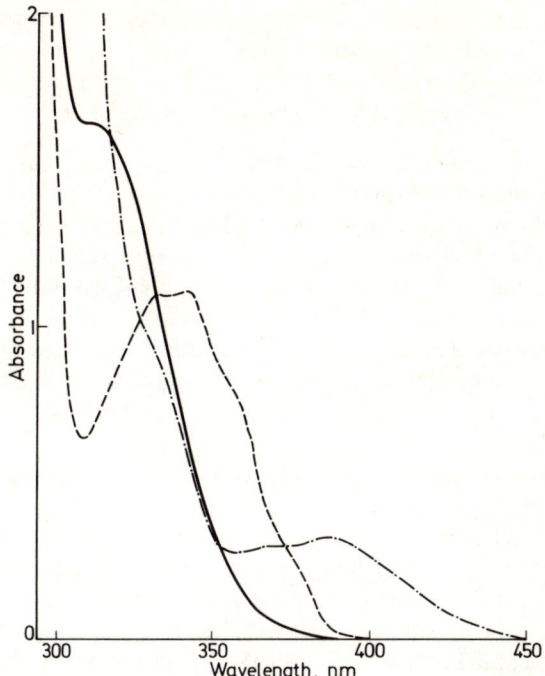

Fig. 1. Ultra-violet absorption spectra of benzoin (———), benzoin methyl ether (– – – –) and benzil (–·–·–) in methyl methacrylate (1.0×10^{-2} M, 5 mm path length)

ether. In methanol solution, the spectra resemble each other much more closely (benzoin: λ_{max} 318 nm, $\log \varepsilon = 2.44$; benzoin methyl ether: λ_{max} 326 nm, $\log \varepsilon = 2.42$); hydrogen bonding between methanol and the carbonyl group of benzoin methyl ether resulting in the characteristic blue shift of the n—π^* transition. The small *red* shift observed in the case of benzoin (λ_{max} 313 nm in methyl methacrylate, 318 nm in methanol) is an interesting effect and may be a result of *inter*-molecular hydrogen bonding disrupting the *intra*-molecular hydrogen bonded structure; acetoin has been reported to exhibit similar behaviour (23).

b) Quenching of Excited States. Experiments involving triplet state quenchers, e.g. piperylene, led Heine *et al.* (7, 19) to the conclusion that whereas benzoin ethers fragment from singlet-excited states, benzoin itself fragments mainly from the triplet state. This could result in the latter exhibiting higher quantum efficiencies for fragmentation in monomers having high triplet energies than in low triplet energy monomers (e.g. styrene, $E_T = 61$ kcal) (24) where triplet-triplet energy transfer may compete with fragmentation of the excited benzoin.[1]

c) Reactivities of Primary Radicals. Norrish Type I cleavage (1) of benzoin, benzoin alkyl ethers and α-alkylated benzoins will in all cases give rise to benzoyl radicals. There is evidence (25) that the unpaired electron of the benzoyl radical is not delocalised into the aromatic nucleus, benzoyl radicals are therefore expected to be reactive species and efficient initiators. The fate of the more stable substituted benzyl radicals $Ph\dot{C}(R_1)$—OR_2, is less certain. For $R_2 = H$, initiation by hydrogen transfer to monomer is a possibility (7), and has been proposed by Braun (26) for initiation of vinyl polymerization by semibenzopinacol radicals, $Ar_2\dot{C}$—OH, generated by thermolysis of benzopinacols. However the comparable efficiencies of benzoin and benzoin methyl ether for bulk methyl methacrylate polymerization (20) show that in this system, no marked effects arise from possible differences in reaction pathways open to $Ph\dot{C}HOH$ and $Ph\dot{C}HOMe$ radicals. From the relationships between rate of polymerization and number-average molecular weight of the polymers formed in the benzoin- and benzoin methyl ether-photoinitiated methyl methacrylate polymerizations it was concluded (20) that primary radical termination of polymer chains did not occur to any significant extent. This may be taken as indirect evidence that both $Ph\dot{C}HOH$ and $Ph\dot{C}HOMe$ radicals do in fact initiate polymerization of this monomer. Earlier reports (Refs. II, IIa) of incorporation of several benzoin and benzoin methyl ether units per polymer molecule remain unexplained.

d) Mobility of Reaction Medium. Differences in mobility between the polyester/styrene and bulk methyl methacrylate systems may influence the relative photoinitiating efficiencies of benzoin derivatives. Cage reactions, e.g. recombination of primary radicals, would be expected to be more important in the former system. Primary radicals from benzoin could also undergo cage reaction to form benzaldehyde (18) whereas an

---

[1] *Note Added in Proof.* It has recently been reported that 4,4'-dimethylamino benzophenone sensitises fragmentation of benzoin isopropyl ether, providing the first example of low energy sensitisation of Type I cleavage. This could be of considerable value in increasing the long wavelength sensitivity of these photoinitiating systems (S. P. Pappas, A. Chattopadhyay, J. Am. Chem. Soc. **95**, 6484 (1973)).

analogous reaction between radicals from benzoin ethers is much less likely viz:

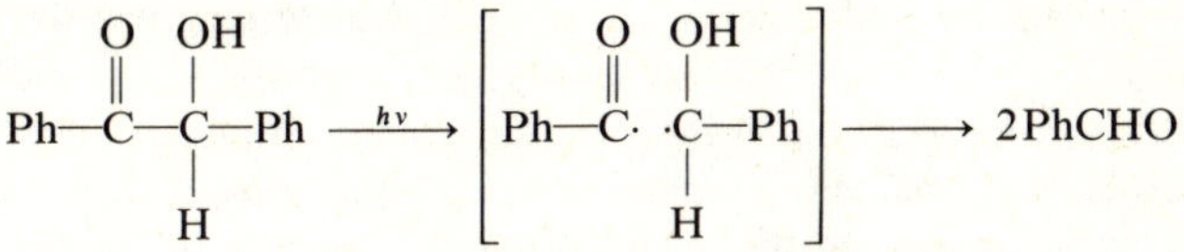

$$Ph-\underset{H}{\underset{|}{\overset{O}{\overset{\|}{C}}}}-\overset{OH}{\overset{|}{C}}-Ph \xrightarrow{h\nu} \left[Ph-\overset{O}{\overset{\|}{C}}\cdot \quad \cdot\underset{H}{\underset{|}{\overset{OH}{\overset{|}{C}}}}-Ph \right] \longrightarrow 2\,PhCHO$$

It is also possible that in less mobile systems, initiation by photoexcited benzoin derivatives takes place by hydrogen abstraction from the environment, as well as by fragmentation. Photoexcited desoxybenzoin ($PhCOCH_2Ph$) is known to take part in hydrogen abstraction reactions (27), and evidence for the hydrogen abstracting capability of benzoin and benzoin methyl ether has been provided by the spin trapping technique (18) referred to earlier. A bifunctional phenoxynitrone scavenger was used. Oxygen-centred radicals and triplet excited ketones are known to abstract the phenolic hydrogen atom with formation of the corresponding phenoxy radical; whereas carbon-centred radicals add on to the scavenger to form stable nitroxide radicals e.g.

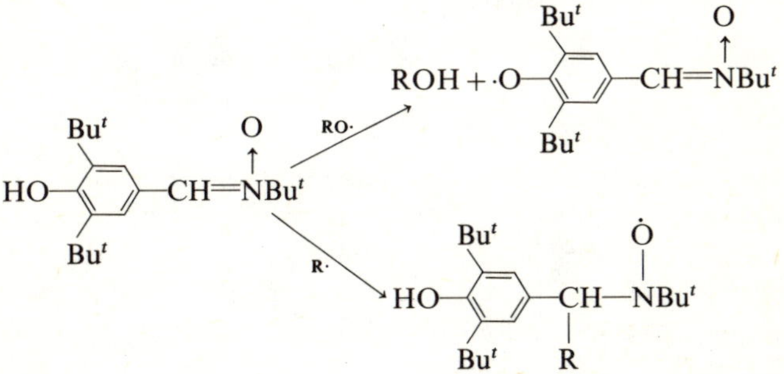

Photolysis of benzoin and benzoin methyl ether in the presence of the phenoxynitrone gave rise to e.s.r. signals from both phenoxy and nitroxide radicals. Hydrogen abstraction by benzoin and benzoin ethers would yield radicals of the type

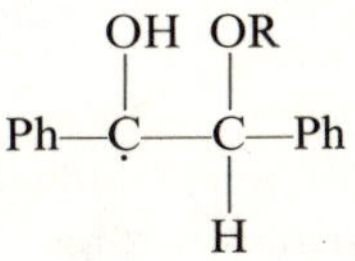

$$Ph-\underset{H}{\underset{|}{\overset{OH}{\overset{|}{C}}}}-\overset{OR}{\overset{|}{C}}-Ph$$

(R = H, alkyl) and there is clear evidence for differing tendencies to terminate polymer chains by the closely related semi-pinacol radicals, $Ar_2\dot{C}$—OH. (This is discussed fully in a later section). Thus it may be that the benzoin-derived radical is more efficient in terminating polymer chains than the alkyl-substituted analogues, perhaps for steric reasons, and that this contributes to the apparent lower efficiency of benzoin in photo-induced crosslinking processes. In the more mobile methyl methacrylate system, addition of 10% v/v tetrahydrofuran (a good hydrogen donor) had no appreciable effect on the rates of benzoin and benzoin methyl ether photoinitiated polymerizations (20), and it may be assumed that under these conditions initiation by hydrogen abstraction is unimportant.

B. Sulphur- and Halogen-Containing Aromatic Carbonyl Compounds

1. β-Ketosulphides

Petropoulos (28) carried out a kinetic study of the photoinitiation of bulk tetraethylene glycol dimethacrylate polymerization by desyl aryl sulphides.

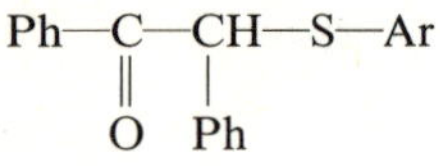

$$\text{Ph—C—CH—S—Ar} \quad \text{Ar = phenyl, o-tolyl, p-tolyl,} $$
$$\underset{\underset{\text{O}}{\parallel}}{}\underset{\text{Ph}}{} \quad \text{p-anisyl, β-naphthyl}$$

Polymerization rates were approximately half-order with respect to photoinitiator concentration and first-order in monomer. A fragmentation mode for desyl aryl sulphides had previously been proposed by Schönberg et al. (29) who isolated bidesyl after prolonged exposure to sunlight of benzene solutions of these compounds:

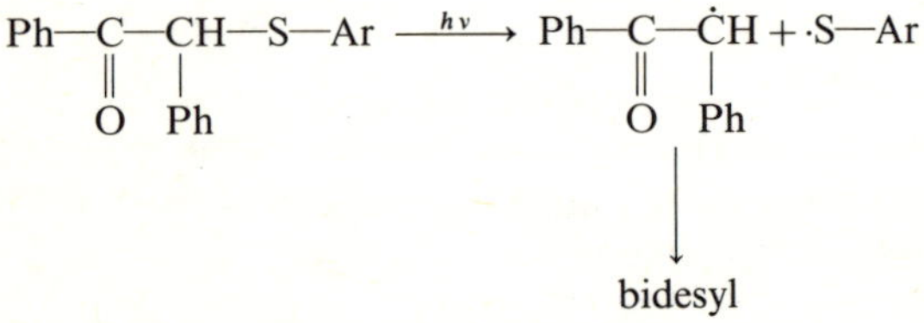

$$\text{Ph—C(=O)—CH(Ph)—S—Ar} \xrightarrow{h\nu} \text{Ph—C(=O)—}\dot{\text{C}}\text{H(Ph)} + \cdot\text{S—Ar}$$

$$\downarrow$$

bidesyl

Petropoulos (28) observed that the ring-substituted compounds (see above) were more efficient photoinitiators than desyl phenyl sulphide itself, and suggested that resonance stabilisation of the aryl thio radical might be important in increasing the rates of dissociation.

Phenyl phenacyl sulphide has also been reported to be a good photoinitiator for the polymerization of methyl methacrylate, acrylonitrile, styrene, and vinyl acetate (30).

$$Ph-\underset{O}{\underset{\|}{C}}-CH_2-S-Ph \xrightarrow{h\nu} Ph-\underset{O}{\underset{\|}{C}}-\dot{C}H_2 + \cdot S-Ph$$

In contrast, ethyl phenacyl sulphide was found to be much less efficient, and an alternative mode of photodecomposition of this compound was suggested (30).

Heine (7) has given an account of studies of the photoinitiating efficiencies of a series of β-ketosulphides and those p-substituted benzophenone sulphides which exhibit "phenylogous β-cleavage".

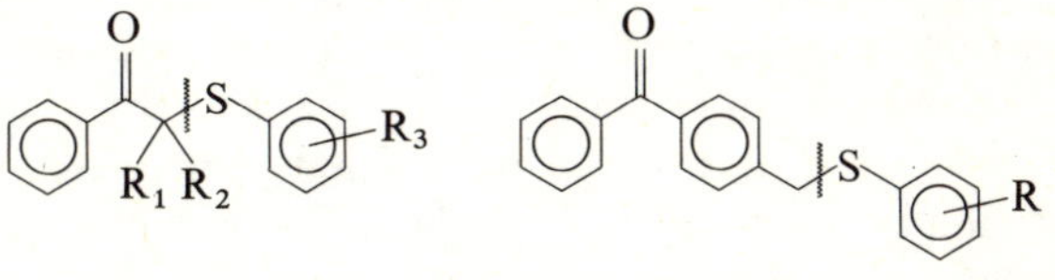

Fragmentation was not suppressed by triplet quenchers such as 1.3-cyclohexadiene or piperylene and the thio aryl radicals were thought to be mainly responsible for initiation.

2. Dibenzoyl Disulphides

p,p′-substituted dibenzoyl disulphides (III) have been reported to act as photoinitiators for the polymerization of styrene and methyl methacrylate, whilst being ineffective as thermal initiators (31).

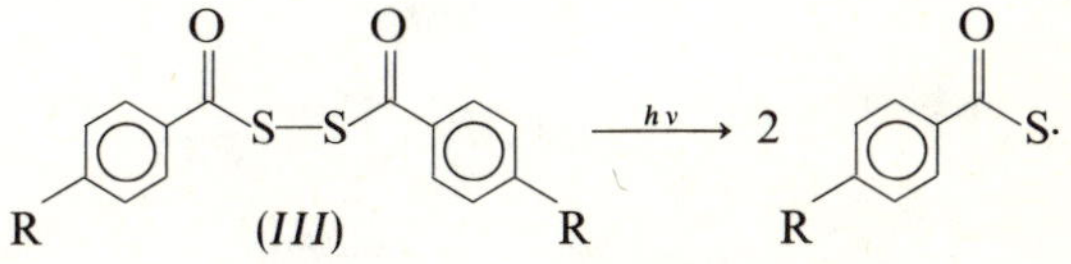

R = Br, Cl, OMe, Me, NO_2, CN.

Polymerization rates were not proportional to the square root of initiator concentration, from which it was concluded that the radicals formed both initiated and terminated polymer chains.

3. S,S'-Diphenyldithiocarbonate

As with the previous compounds (*III*), photodecomposition of S,S'-diphenyldithiocarbonate (*IV*) gives rise to two identical radicals which, from kinetic evidence, both initiate and terminate the polymerization of styrene and methyl methacrylate (*32*).

$$\text{Ph—S—}\underset{\underset{O}{\|}}{C}\text{—S—Ph} \xrightarrow{h\nu} 2\,\text{Ph—S·} + CO$$

(*IV*)

At low initiator concentrations, the normal square-root dependence of polymerization rate on initiator concentration was observed, but deviations became apparent at higher concentrations. An alternative explanation for all these deviations could be that at the higher concentrations of photoinitiators nearly all of the incident light was being absorbed in which case the polymerization rates would tend to become independent of photoinitiator concentrations. Moreover as the latter are increased further, a decrease in overall polymerization rate may be observed (*20*) due to initiating radicals being formed almost exclusively in a very narrow region nearest to the light source.

4. Halides

Halogen atoms are well-known initiating species (*33*) and can be readily generated by photolysis of a wide range of suitably substituted phenyl ketones. Examples of those which have been reported to be efficient photoinitiators are given below (*7*).

phenyl α-haloketones

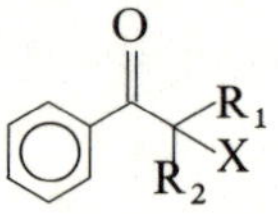

$X = Cl, Br.$
$R_1, R_2 = Cl, Br, H,$ alkyl, phenyl.

ω-trichlorocrotophenone

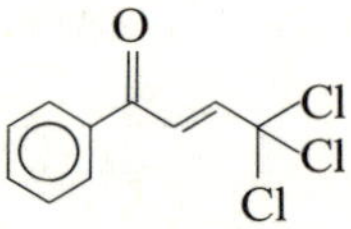

halogenated p-alkyl benzophenones

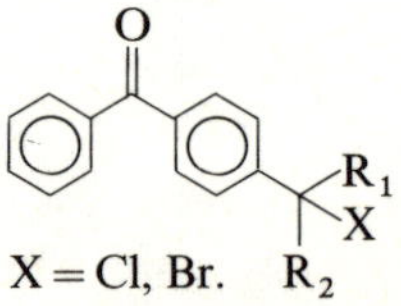

$X = Cl, Br.$
$R_1, R_2 = Cl, Br, H,$ alkyl.

.The part played by the more stable fragments resulting from carbon-halogen cleavage is again uncertain, but Heine (7) has indicated a reduced tendency for these radicals to add to unsaturated systems. Dimeric products have been isolated in the case of photolyses carried out in benzene, e.g. (7).

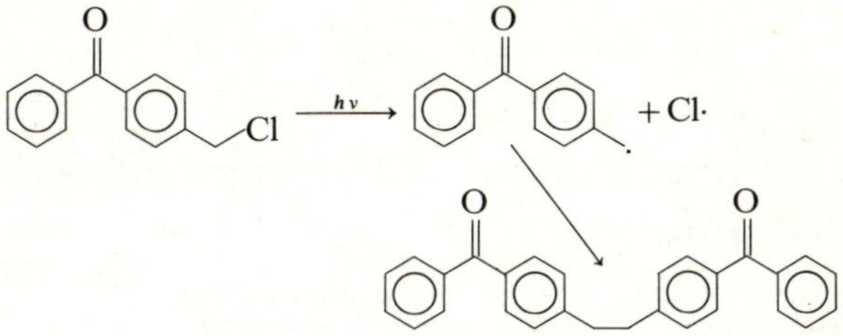

In polymerizing systems, the resonance-stabilised fragments could act mainly as terminating species, but it remains a fact nonetheless, that their formation, simultaneous with a reactive, initiating radical, is a general feature of the known, efficient, photoinitiator systems; presumably because their relative stability facilitates dissociation. This may help to explain why benzil

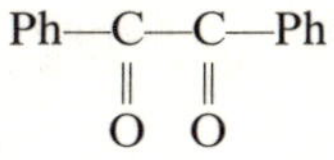

$$Ph-\underset{\underset{O}{\|}}{C}-\underset{\underset{O}{\|}}{C}-Ph$$

contrary to previous assumptions (11, 16), fragments very inefficiently, if at all, on exposure to ultra-violet light (34), since dissociation into two comparatively unstable benzoyl radicals would be unfavourable. Photoinitiation by benzil is now thought to proceed by a hydrogen-abstraction mechanism (20), and will be discussed more fully in the following section.

III. Photoinduced Hydrogen Abstraction by Aromatic Carbonyl Compounds

A. General Mechanism

The best known hydrogen abstraction reaction of aromatic carbonyl compounds is that involved in the formation of benzopinacol on ultra-violet irradiation of solutions of benzophenone in ethers and alcohols.

This has been the subject of a large of number of studies (*35—41*) and quantum yields for photoreduction approach a value of 2, (*41*) i.e. ~2 molecules of benzophenone consumed per quantum of light absorbed. The reaction is generally considered to proceed via free radical intermediates according to the scheme shown below (*38*) (for photoreduction of benzophenone in isopropanol).

$$Ph_2C=O \xrightarrow{h\nu} Ph_2C=O^*$$

$$Ph_2C=O^* + Me_2CHOH \longrightarrow Ph_2\dot{C}-OH + Me_2\dot{C}-OH$$

$$Me_2\dot{C}-OH + Ph_2C=O \longrightarrow Me_2C=O + Ph_2\dot{C}-OH$$

$$2\,Ph_2\dot{C}-OH \longrightarrow \underset{\underset{OH}{|}}{Ph_2C}-\underset{\underset{OH}{|}}{CPh_2}$$

Photoreduction of benzophenone by isopropanol is accompanied by the formation of a coloured compound which Filipescu and Minn (*39*) proposed was an adduct (see below) of the radicals resulting from hydrogen abstraction. These authors further proposed that this adduct was a key intermediate in the photoreduction, semipinacol radicals being generated by its reactions with ground state benzophenone.

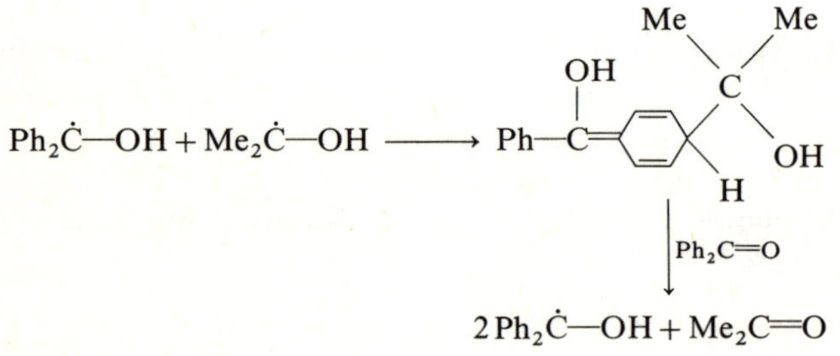

$$Ph_2\dot{C}-OH + Me_2\dot{C}-OH \longrightarrow \text{[adduct V]} \qquad (V)$$

$$\downarrow Ph_2C=O$$

$$2\,Ph_2\dot{C}-OH + Me_2C=O$$

The precise reaction mechanism is still a matter of controversy. Weiner (*40*) has pointed out that the proposition that the coloured compound (*V*) is a key intermediate cannot be reconciled with available kinetic data, and considers its formation to be incidental to the production of benzopinacol. Weiner (*40*) has also reported evidence for formation of a "direct" adduct, $Ph_2C(OH)C(OH)Me_2$, which apparently arises from an in-cage reaction accounting for approximately 10% of the radical intermediates.

Initial formation of radicals by photoinduced hydrogen abstraction from the substrate is not in dispute and, as noted in Section I, hydrogen abstraction from alcohols, ethers and hydrocarbons is effected with high quantum efficiencies by aromatic carbonyl compounds in which the lowest lying triplet level is of n—π^* type, e.g. benzophenone, anthraquinone (42—44).

B. Photoinitiation by Benzophenone Derivatives

Although benzophenone derivatives have found application in a number of photopolymerization processes (10), these have usually involved cross-linking reactions induced by energy transfer from photoexcited benzophenones to groups (e.g. cinnamate) attached to the polymer chains. In such cases, the benzophenones act as sensitisers and not as photoinitiators. An example in which hydrogen abstraction by photoexcited benzophenone has been utilized for photoinitiation is the ultra-violet induced grafting of styrene on to polyethylene (8, 45). Recently, extensive studies concerned with photoinitiation of vinyl polymerization by benzophenone derivatives in homogeneous media have been carried out, and are described in detail below.

1. Kinetics and Mechanism of Methyl Methacrylate Polymerization Photoinitiated by Benzophenones in Tetrahydrofuran

Block, Ledwith and Taylor (46) have reported the results of a detailed investigation into the efficiencies of benzophenone, 3,3′,4,4′-benzophenone tetracarboxylic dianhydride (BTDA) and 3,3′,4,4′-tetramethoxycarbonyl benzophenone (TMCB) as photoinitiators for the polymerization of methyl methacrylate in tetrahydrofuran at 30° C *in vacuo*.

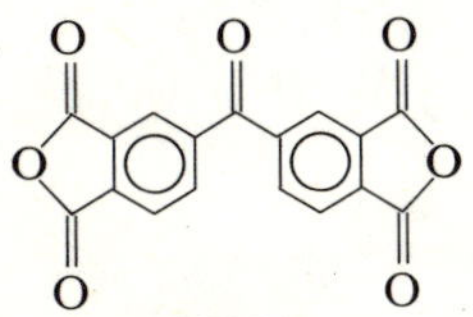

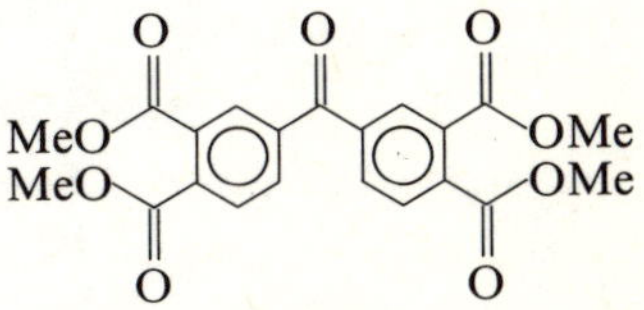

Efficiencies increased in the order benzophenone < TCMB < BTDA, although the ultra-violet absorption spectra of these compounds showed only minor variations in the region 330—380 nm. The kinetic results were consistent with the expected free radical mechanism, polymerization rates being half-order with respect to the concentrations of the benzophenones. However the viscosity-average molecular weights of the resulting polymers were considerably less than those measured for polymers formed at comparable rates, at the same temperature, using azobisisobutyronitrile as photoinitiator (again in tetrahydrofuran solution). The effect was most pronounced in the case of polymerizations photoinitiated by benzophenone itself, and indicated the participation of an extra termination or transfer step. Addition of benzophenone, benzhydrol, and benzopinacol to thermal polymerizations initiated by azobisisobutyronitrile had no effect on the rate of polymerization or the molecular weight of the polymer formed, and it was concluded that the reduction in molecular weight of polymers formed in the reactions photoinitiated by benzophenones was caused by an intermediate formed in the initiation process. The presence of a solvent with readily abstractable hydrogen atoms was shown to be essential for efficient photoinitiation by benzophenone; in benzene (a poor hydrogen donor) the rate of photo-induced polymerization of methyl methacrylate was only marginally increased by the presence of benzophenone. The experimental observations were consistent with the reaction scheme shown below.

Initiation: $Ar_2C=O \xrightarrow{h\nu} [Ar_2C=O]S_1 \underset{n-\pi^*}{} \to [Ar_2C=O]T_1 \underset{n-\pi^*}{}$

$[Ar_2C=O]T_1 \underset{n-\pi^*}{} + THF \to Ar_2\dot{C}-OH + THF\cdot$

$THF\cdot + M \to P_1^{\cdot}$

Propagation: $P_n^{\cdot} + M \to P_{n+1}^{\cdot}$

Termination: $P_m^{\cdot} + P_n^{\cdot} \xrightarrow{k_t} Polymer$

$Ar_2\dot{C}-OH + Ar_2\dot{C}-OH \xrightarrow{k_c} Ar_2\underset{OH}{C}-\underset{OH}{C}Ar_2$

$Ar_2\dot{C}-OH + P_n^{\cdot} \xrightarrow{k_t'} Polymer$

Relative efficiencies of the three benzophenone derivatives, and the sequence observed in the polymer molecular weight/polymerization rate relationships, could be due to differing tendencies of the respective semi-pinacol radicals to undergo self-combination, since it seems reasonable to assume that semi-pinacol radicals should show rather small differences in their abilities to combine with growing polymer radicals. However it was pointed out that this termination reaction may involve hydrogen transfer rather than direct combination i.e.

$$Ar_2\dot{C}-OH + P\cdot \rightarrow Ar_2C=O + PH$$

and hence differences in the efficiency of hydrogen transfer from the semi-pinacol radicals could be an alternative explanation for the experimental observations noted above. More recently, Hammond and his collaborators (47), using a combination of flash photolytic and e.s.r. techniques, have investigated, in considerable detail, the rates of recombination of semi-pinacol radicals (p-$XC_6H_4)_2\dot{C}$—OH where X = H, OMe, Cl. A most important conclusion from these studies is that, while in purely alkane solvents rates of recombination of the semi-pinacol radicals were essentially independent of the nature of substituents (X) i.e. essentially diffusion controlled, the corresponding reactions in solvents such as benzene or isopropanol exhibit a pronounced effect of substituent (X) on reaction rate. Clearly then, relative rates of self-combination of semi-pinacol radicals, or of combination with growing polymer radicals (termination), could be different according to the nature of substituents and any specific or general solvation forces between solvent, monomer and the reacting radicals.

The inclusion of a primary radical termination process in radical polymerization schemes usually leads to kinetic equations which cannot be reduced to a straightforward expression for the orders of reaction with respect to initiator and monomer concentration (48). It is interesting to note therefore that, using only the normal approximations, such an expression can be derived from the above scheme which predicts that the polymerization will be half-order in initiator and light intensity and first-order in monomer concentration, *despite* the participation of primary radical termination. Straightforward solution of the kinetics is made possible by the following assumptions, implicit in the scheme:

(i) all of the solvent-derived radicals initiate polymerization,
(ii) initiation by the semi-pinacol radicals is negligible.

These lead to a simple relationship between the stationary concentrations of semi-pinacol radicals and propagating polymer radicals, i.e.

$$k_t[P\cdot]^2 = k_c[Ar_2\dot{C}-OH]^2.$$

Radicals formed by hydrogen abstraction from cyclic ethers are known to add readily to olefins (49), since α-substituted cyclic ethers can be prepared in good yield from olefins which do not undergo radical homopolymerization, e.g. maleic anhydride (*50*), 1-octene (*51*).

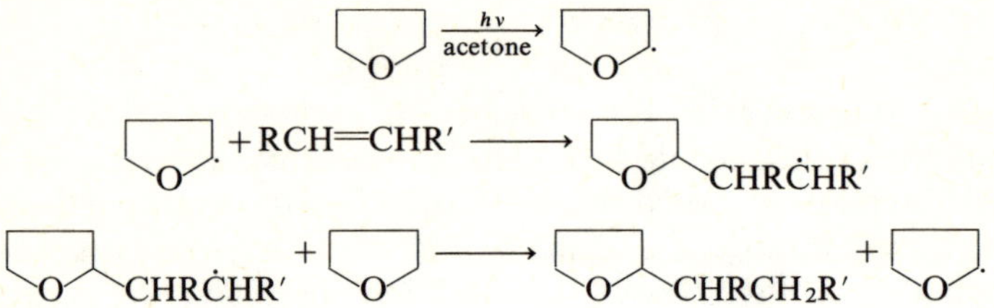

$$\text{THF} \xrightarrow[\text{acetone}]{h\nu} \text{THF}\cdot$$

$$\text{THF}\cdot + \text{RCH}=\text{CHR}' \longrightarrow \text{THF-CHRĊHR}'$$

$$\text{THF-CHRĊHR}' + \text{THF} \longrightarrow \text{THF-CHRCH}_2\text{R}' + \text{THF}\cdot$$

Thus it seems reasonable to assume that in the presence of a polymerizable olefin such as methyl methacrylate, tetrahydrofuran-derived radicals will initiate with high efficiency.

The validity of assumption (ii) above is less certain. Braun *et al.* (*26*) have reported that semi-pinacol radicals can initiate the polymerization of methyl methacrylate and styrene at temperatures above 40° C. The semi-pinacol radicals were generated by thermolysis of aromatic pinacols, and kinetic studies indicated that the initiation mechanism involved transfer of a hydrogen atom from the semi-pinacol radical to monomer, i.e.

$$\underset{\underset{\text{OH}}{|}}{\text{Ph}_2\text{C}}\text{—}\underset{\underset{\text{OH}}{|}}{\text{CPh}_2} \xrightarrow{\Delta} 2\text{Ph}_2\dot{\text{C}}\text{—OH}$$

$$\text{Ph}_2\dot{\text{C}}\text{—OH} + \text{CH}_2=\text{CHR} \longrightarrow \text{Ph}_2\text{C}=\text{O} + \text{CH}_3\text{—}\dot{\text{C}}\text{HR}$$

In order to assess the importance of initiation by semi-pinacol radicals at a lower temperature, Hutchison *et al.* (*52*) compared rates of polymerization observed on photolysis of methyl methacrylate/benzene mixtures containing benzophenone ($\sim 5 \times 10^{-3}$ M) alone, and benzophenone + benzhydrol (~ 0.15 M). The aim was to encourage increased semi-pinacol radical formation in the latter mixtures, since it was anticipated that photoexcited benzophenone would hydrogen-abstract from benzhydrol (rather than methyl methacrylate or benzene) thus giving rise to semi-pinacol radicals from *both* precursors:

$$\text{Ph}_2\text{C}=\text{O}(T_1) + \text{Ph}_2\text{CHOH} \longrightarrow 2\text{Ph}_2\dot{\text{C}}\text{—OH}$$

At 30° C, the presence of benzhydrol retarded the polymerization rate by a factor of approximately 2, and also caused the (weight-) average molecular weight of the resulting polymer to be drastically reduced (from 340000 to 60000). In the case of polymerizations photoinitiated by azobisisobutyronitrile, the presence of benzhydrol affected neither the rate of polymerization nor the average molecular weight of the polymer produced, demonstrating conclusively that semi-pinacol radicals act as terminating species in the polymerization of methyl methacrylate at this temperature. The possibility that semi-pinacol radicals also initiate polymerization to some extent at 30° C could not be completely excluded, but was shown to be of minor importance.

Another factor which could influence the photoinitiating efficiencies of benzophenone derivatives is the extent of in-cage reaction of the primary radicals resulting from hydrogen abstraction. Cage reactions reported to occur in the photoreduction of benzophenone by isopropanol have already been described (Section III. A). Photolysis of benzophenone in tetrahydrofuran was found (52) to cause an increase in the ultra-violet absorption spectrum between 300 and 370 nm. This was ascribed to the formation of an adduct arising from combination of semi-pinacol and tetrahydrofuran-derived radicals, and analogous in structure to that proposed by Filipescu and Minn (39) (V, see Section III. A). The presence of increasing proportions of methyl methacrylate (in the range 0—50% by volume) resulted in a corresponding decrease in the rate of formation of the adduct. Two possible mechanistic interpretations of this were offered.

(i) The quantum yield for production of the primary radicals is not altered by addition of methyl methacrylate, but the proportion of tetrahydrofuran-derived radicals which react with methyl methacrylate increases with increasing concentration of the latter, thus the rate of adduct formation is decreased.

(ii) The proportion of primary radicals which combine to give adduct remains constant, but the overall rate of primary radical formation decreases with increasing methyl methacrylate concentration owing to quenching of triplet excited benzophenone by methyl methacrylate.

In the benzophenone-photoinitiated polymerization of methyl methacrylate, interpretation (i) would tend to *increase* the order of reaction with respect to monomer concentration, whereas interpretation (ii) would tend to *decrease* this, assuming that the species formed as the result of quenching initiated polymerization less efficiently than the solvent-derived radicals. The data of Block *et al.* (46) give a value of 0.84 for the order in monomer, and it could be argued that this is significantly lower than the value of unity predicted by the suggested kinetic scheme and favours interpretation (ii). Values obtained for monomer orders in the case of the BTDA and TCMB photoinitiated polymerizations were much closer to 1 (1.02 and 0.98 respectively) which, following the present reasoning, would indicate that the excited states of these derivatives are not quenched by methyl methacrylate.

Thus the relative photoinitiating efficiencies of benzophenone derivatives may be influenced by whether or not their photoexcited states are quenched by monomer, and the fraction of in-cage combination of the radicals formed by hydrogen abstraction, as well as the extents to which the respective semi-pinacol radicals terminate polymer chains. At elevated temperatures, differences in the initiating abilities of the semi-pinacol radicals must also be considered.

2. Other Monomers and Solvents

Heine (7) has reported that benzophenone is practically inactive as a photoinitiator in polymerizable mixtures containing styrene, and that this is because the latter quenches triplet-excited benzophenone. However Block *et al.* (46) observed that photoinitiation of styrene polymerization was effected to some extent by benzophenone, and by TCMB, in the presence of tetrahydrofuran.

The photo-induced polymerization of acrylonitrile, again in tetrahydrofuran solution, is greatly enhanced by the presence of benzophenone (46), which in this case exhibits a higher efficiency than TCMB. Both these compounds *retarded* the photoinduced polymerization of vinyl acetate in tetrahydrofuran, but did act as photoinitiators when toluene was used as solvent (46), possibly reflecting the differing abilities of the solvent-derived radicals to add to vinyl acetate.

Solvent effectiveness in the benzophenone-photoinitiated polymerization of methyl methacrylate was in the order tetrahydrofuran > isopropanol > toluene > benzene (46). In the case of TCMB, isopropanol gave slightly higher polymerization rates than tetrahydrofuran, but both were again considerably more effective than toluene and benzene. As might be expected, monomers which themselves contain ether groups, e.g. diethylene glycol diacrylate (7), do not require addition of a separate hydrogen-donor solvent for efficient photoinitiation by benzophenone.

3. Intramolecular Hydrogen Abstraction

The requirement that a substrate containing labile hydrogen atoms must be present for efficient photoinitiation by benzophenone derivatives limits the applicability of these systems. One possible way of avoiding the need for hydrogen-donating additives is indicated in apparently unrelated work described by Breslow (53, 54), involving photolysis of benzophenone derivatives containing groups large enough to enable intramolecular hydrogen abstraction to take place. The original work of Breslow and his collaborators was primarily concerned with selective functionalisation of the steroid skeleton by photolysis of compounds in which this was attached to the benzophenone moiety. Preliminary work was carried out on long-chain n-alkyl esters of benzophenone-4-carboxylic acid, and the cyclised products obtained indicated the intermediacy of diradicals formed by intramolecular hydrogen abstraction, e.g.

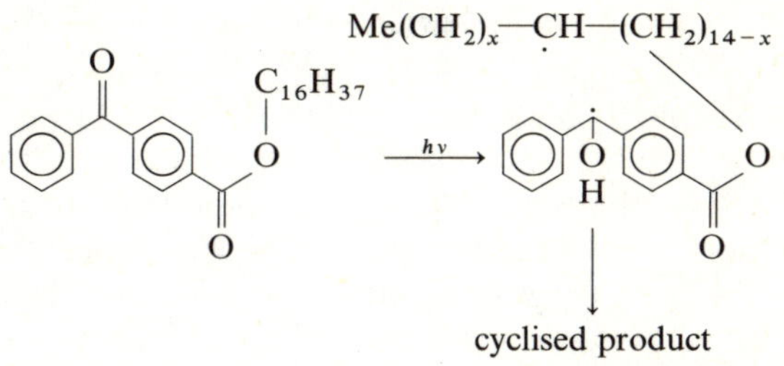

cyclised product

The feasibility of utilising similar intramolecular hydrogen abstraction reactions for photoinitiation was therefore tested (55). Reactions of 3,3',4,4'-benzophenone tetracarboxylic dianhydride (BTDA) with primary n-alkyl amines readily yield a series of N-substituted imides.

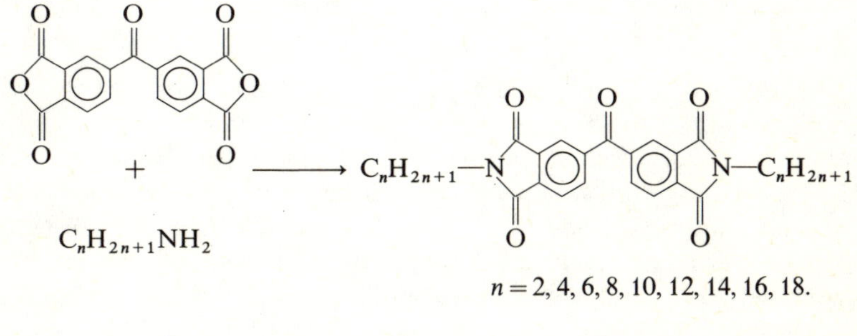

$n = 2, 4, 6, 8, 10, 12, 14, 16, 18$.

The short-chain compounds (C_2—C_{10}), for which space filling models suggested that intramolecular hydrogen abstraction was not possible, were found to be significantly more efficient than benzophenone itself as photoinitiators for the polymerization of bulk methyl methacrylate (at 30° C, *in vacuo*, photoinitiator concentrations 10^{-3} M). The efficiencies of the C_{12}-, C_{14}- and C_{16}-imides were slightly higher and comparable with that of BTDA, whilst the C_{18}-imide exhibited the highest efficiency of all. However even the latter gave an approximately three-fold increase in polymerization rate on addition of 10% v/v tetrahydrofuran to the system, showing clearly that the full potential for initiation by intramolecular hydrogen abstraction was far from being attained. It seemed likely therefore that any intramolecular hydrogen abstraction which does take place leads predominantly to cyclised products, analogous to those observed by Breslow, rather than to initiation of polymer chains[2].

C. Photoinitiation by Benzil

Benzil has frequently been used as a means of generating free radicals in polymerization systems subjected to ultra-violet irradiation (*11*, *16*, *56—58*). In studies of the benzil-photoinitiated polymerizations of methyl methacrylate, and vinyl acetate, Melville (*16*) assumed that initiation was brought about by fragmentation of photoexcited benzil into two benzoyl radicals. However a survey of the photochemistry of benzil (*34*) indicates that such a cleavage does not in fact take place in solution; studies of the products formed on irradiation of benzil in cyclohexane (*59*), cumene and isopropanol (*60*) can be rationalised on the basis of initial hydrogen abstraction from solvent by photoexcited benzil, e.g.

$$\text{Ph}-\underset{\underset{O}{\|}}{C}-\underset{\underset{O}{\|}}{C}-\text{Ph} \xrightarrow[C_6H_{12}]{h\nu} \text{Ph}-\underset{\underset{O}{\|}}{C}-\underset{\underset{OH}{|}}{\dot{C}}-\text{Ph} + \overset{\cdot}{\bigcirc}$$

The results of spin-trapping experiments (*18*) also support this view; no radicals were trapped on photolysis of benzil in benzene solutions of compounds I and II (see Section II.A.1) but in methanol solutions of II, $\cdot CH_2OH$ radicals were scavenged, indicating that hydrogen abstraction by photoexcited benzil had occurred (*18*). Thus the behaviour of benzil in these experiments contrasted markedly with that of benzoin and benzoin methyl ether, described previously.

---

[2] Note Added in Proof. However the photochemistry of the above N-substituted imides may be rather more complex than was first anticipated, since a recent publication indicates that hydrogen abstraction and subsequent cyclisation may also be brought about by carbonyl groups associated with the imide function (Y. Kanaoka, Y. Migita, and K. Koyama, Tet. Letters, 1193 (1973)).

As a photoinitiator for the polymerization of bulk methyl methacrylate, benzil was found to be considerably less efficient than benzoin or benzoin methyl ether (20), e.g. at photoinitiator concentrations of 10^{-2} M, the polymerization rate observed using benzoin was eight times that observed using benzil, despite the fact that the latter was found to absorb three times as much of the incident radiation (Fig. 1). The photoinitiating efficiency of benzil was improved by a factor of three on addition to the methyl methacrylate of 10% v/v tetrahydrofuran, whereas the same additive had no appreciable effect on rates of benzoin- and benzoin methyl ether-photoinitiated polymerizations; direct evidence that photoinitiation by benzil proceeds by a hydrogen abstraction mechanism rather than by fragmentation.

D. Photoinitiation by Quinones

Quinones are a special case of aromatic carbonyl compound and, in general, give lowest lying excited states having (n, π^*) character. This means that photoinduced hydrogen abstraction processes are as common for triplet quinones as for benzophenone derivatives. Anthraquinone (42—44) and its derivatives (61—65) have been most studied from this point of view but there are several reports of photoreduction of other quinones (66—68). Generally speaking the types of hydrogen donor molecules (alcohols, ethers) effective for photoreduction of quinones are the same as for benzophenones and reaction mechanisms are very similar. One difference between quinones and benzophenones is that the former more readily yield anion radicals on reduction, particularly in alkaline media e.g.

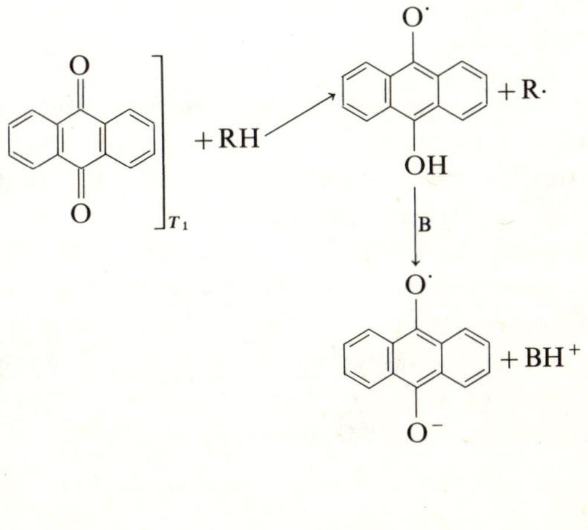

A further point of difference between photoreactions of quinones and benzophenones is the much greater and often useful involvement of intermediate peroxy compounds in reactions of the former conducted in air (69—71).

1. Photoinitiation of Methyl Methacrylate Polymerization by Anthraquinone in Tetrahydrofuran

There are numerous references in the Patent Literature (72—76) to the use of anthraquinone and similar compounds as photoinitiators of polymerization or crosslinking required in the preparation of printing plates etc., although apparently little is known of the detailed reaction mechanisms. One literature report (77) described the polymerization of aqueous methyl acrylate by sodium anthraquinone-2-sulphonate in the presence of chloride ions, with a conclusion that the initiating species are chlorine atoms.

A comprehensive study of the free radical polymerization of methyl methacrylate (MMA) photoinitiated by anthraquinone and 2-tert-butyl anthraquinone in tetrahydrofuran has been completed by Ledwith, Ndaalio and Taylor (78). 2-tert-butyl anthraquinone is frequently used in applications (72—76) on account of its greater solubility and compatility with solvents and polymers but detailed kinetic and mechanism studies (78) show that anthraquinone and 2-tert-butyl anthraquinone behave in an almost identical manner. Consequently results for the former quinone only are considered here.

Following conventional dilatometric studies in high vacuum systems, polymerization of MMA photoinduced by anthraquinone (AQ) in THF at 30° C was shown to obey the following rate expression:

$$-\frac{d[\text{MMA}]}{dt} = K[\text{MMA}]^{1.0}[\text{AQ}]^{0.5}[I_0]^{0.5}$$

where I_0 is the incident light intensity. Determination of molecular weights led to an estimate for $k_p/k_t^{1/2} = 0.0429$ litre$^{1/2}$ mole$^{-1/2}$ sec$^{-1/2}$, a value, similar to those obtained with benzophenone initiators (Section III.B.1) but somewhat lower than that (0.0658) for photoinitiation by azobisisobutyronitrile (AIBN) in the same system. Since AQ was shown not to retard the polymerization of MMA thermally initiated by AIBN at 50° C, nor to cause a reduction in the molecular weights of the polymers formed, this lower value of $k_p/k_t^{1/2}$ must represent an additional termination process caused by an intermediate or product of the photoreactions involved in initiation by anthraquinone.

Of particular value in interpreting detailed mechanisms in these systems were the spectral changes noted during polymerization. Irradiation of AQ in THF without monomer gave rise to a blue fluorescing product with absorption maxima at 400 and 374 nm, similar to photo adducts from THF and other quinones (79). In the presence of polymerizable monomer however, the solutions developed a characteristic blue-green fluorescence with an absorption spectrum (λ_{max} 383 nm) identical to that reported (43) for anthrahydroquinone. The latter product rapidly reached a comparatively low, constant concentration during polymerization and both this material, and the adduct formed in high conversion by photolysis of AQ in THF without monomer, were immediately destroyed by exposure to atmospheric oxygen. A reaction scheme consistent with the experimental observations is indicated below:

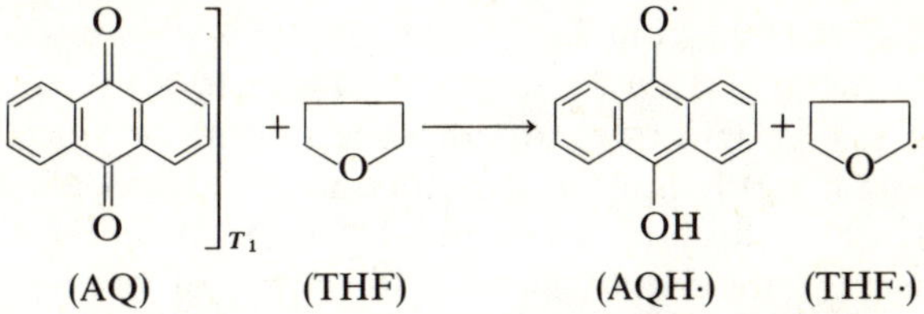

$$[AQ]_{T_1} + THF \longrightarrow AQH\cdot + THF\cdot$$

$$THF\cdot + MMA \longrightarrow \text{Polymer radical } (P\cdot)$$

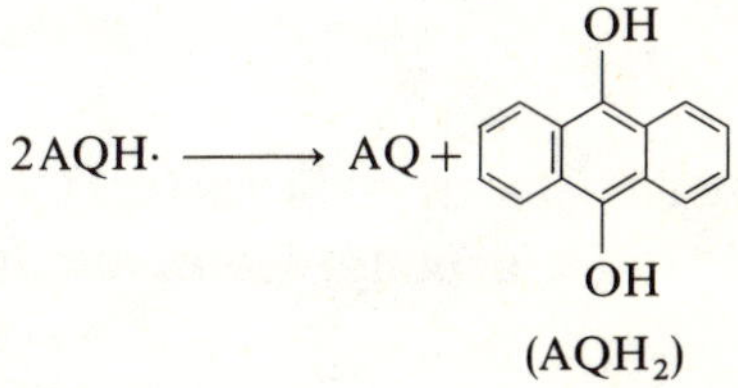

$$2\,AQH\cdot \longrightarrow AQ + AQH_2$$

$$AQH\cdot + P\cdot \longrightarrow \text{Terminated polymer}$$

$$AQH_2 + P\cdot \longrightarrow \text{Terminated polymer} + AQH\cdot$$

$$AQH\cdot + THF\cdot \longrightarrow \text{(adduct)}$$

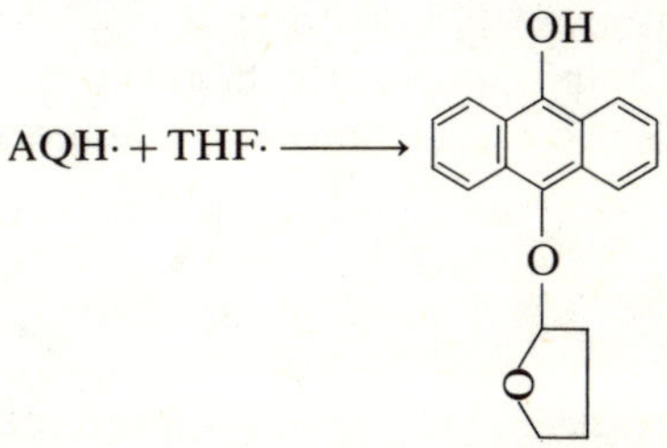

The proposed reaction scheme assumes that the observed steady state concentration of anthrahydroquinone (AQH_2) results from competition between disproportionation of the initially formed semiquinone radical, and termination of polymer chains by AQH· and by anthrahydroquinone (with regeneration of AQH·).

Under similar reaction conditions, other quinones were also found to act as photoinitiators for MMA polymerization in THF with phenanthraquinone being the most useful and approximately twice as efficient as anthraquinone. Solvents other than THF, having hydrogen donor capability, such as alcohols, toluene and cumene were equally useful for MMA polymerization but choice of solvent governs the nature of the initiating radical and hence becomes very important when using monomers other than MMA. Thus photoinitiated polymerization of acrylonitrile and styrene, but not vinyl acetate, could be achieved by the system AQ/THF. On the other hand, toluene/AQ systems were quite efficient for polymerization of vinyl acetate. Inefficient photoinitiation of vinyl acetate polymerization by AQ/THF is not surprising since the initiating radical (derived from THF) is an α-alkoxy alkyl radical similar in structure to those derived by radical addition to alkyl vinyl ethers. The latter are known (80) to retard the free radical polymerization of vinyl acetate and it would appear, therefore, that α-alkoxy alkyl radicals do not add to vinyl acetate. It should be noted however that vinyl acetate does not interfere with the photoreactions of AQ and THF as shown by formation of blue fluorescing adduct.

IV. Photoinduced Electron Transfer Reactions of Aromatic Carbonyl Compounds

A. Reactions Involving N-vinyl Carbazole

Photolysis of solutions of N-vinyl carbazole (NVC) in the presence of aromatic carbonyl compounds such as benzophenone, anthraquinone, fluorenone, and chloranil leads to efficient cyclodimerisation producing the cyclobutane (VI).

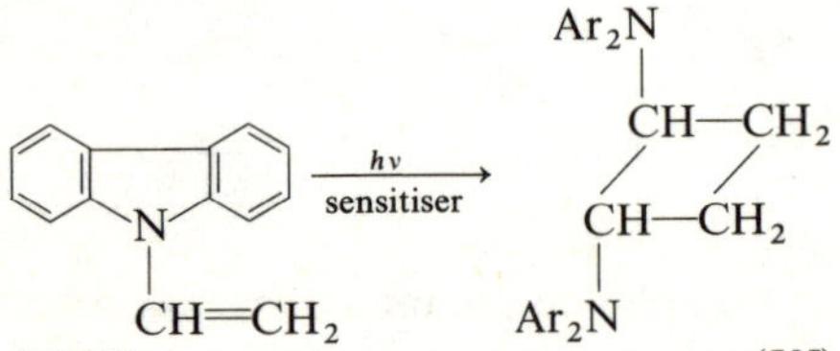

The reaction mechanisms have been discussed in detail (81, 82) but a brief summary would be useful at this stage. Essentially this photocyclodimerisation is quantitative if irradiation is prolonged and of particular interest is the fact that dissolved air or oxygen has a pronounced co-catalytic function without giving oxygen containing products. For many effective sensitisers, energy transfer mechanisms are ruled out because of the relative triplet state energies of sensitisers and NVC and in all cases quantum yields were in excess of unity. A chain reaction mechanism must, therefore, be operative and electron transfer processes for initiation, propagation, and termination have been proposed viz:

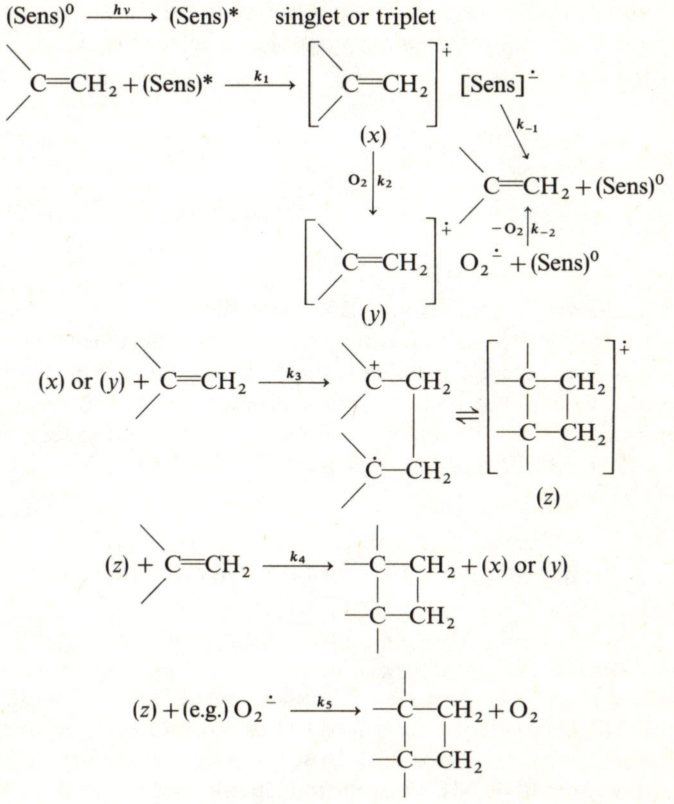

$$(\text{Sens})^0 \xrightarrow{h\nu} (\text{Sens})^* \quad \text{singlet or triplet}$$

$$\underset{}{\diagdown}\!\!\text{C}\!\!=\!\!\text{CH}_2 + (\text{Sens})^* \xrightarrow{k_1} \left[\underset{}{\diagdown}\!\!\text{C}\!\!=\!\!\text{CH}_2 \right]^{\pm} [\text{Sens}]^{\pm}$$
(x)

$$\downarrow O_2 \, k_2 \qquad \xrightarrow{k_{-1}} \underset{}{\diagdown}\!\!\text{C}\!\!=\!\!\text{CH}_2 + (\text{Sens})^0$$

$$\left[\underset{}{\diagdown}\!\!\text{C}\!\!=\!\!\text{CH}_2 \right]^{\pm} \quad \underset{\uparrow -O_2 \, k_{-2}}{} \quad O_2^{\pm} + (\text{Sens})^0$$
(y)

$$(x) \text{ or } (y) + \underset{}{\diagdown}\!\!\text{C}\!\!=\!\!\text{CH}_2 \xrightarrow{k_3} \begin{array}{c} \overset{+}{\text{C}}\!\!-\!\!\text{CH}_2 \\ | \\ \dot{\text{C}}\!\!-\!\!\text{CH}_2 \end{array} \rightleftharpoons \left[\begin{array}{c} -\text{C}\!\!-\!\!\text{CH}_2 \\ | \\ -\text{C}\!\!-\!\!\text{CH}_2 \end{array} \right]^{\pm}$$
(z)

$$(z) + \underset{}{\diagdown}\!\!\text{C}\!\!=\!\!\text{CH}_2 \xrightarrow{k_4} \begin{array}{c} -\text{C}\!\!-\!\!\text{CH}_2 \\ | \\ -\text{C}\!\!-\!\!\text{CH}_2 \end{array} + (x) \text{ or } (y)$$

$$(z) + (\text{e.g.}) O_2^{\pm} \xrightarrow{k_5} \begin{array}{c} -\text{C}\!\!-\!\!\text{CH}_2 \\ | \\ -\text{C}\!\!-\!\!\text{CH}_2 \end{array} + O_2$$

The actual quantum yield observed for any reaction will depend on the relative values of the rate constants indicated in the scheme which, in turn, will depend upon the nature of the sensitiser, its reduction potential, and the interaction of all reacting intermediates with solvent. Exclusion of oxygen from the reaction systems leads, in many cases, to comparatively inefficient free radical polymerization of NVC — this is especially the case for sensitisers such as benzophenone acting via triplet excited states (83). Solvent also has a marked effect on the relative efficiencies of polymerization and cyclodimerisation (84). However for some sensitisers, notably fluorenone, the correlation between fluorescence quenching by NVC and efficiency of cyclodimerisation argues strongly in favour of excited singlet fluorenone as reactant. Cyclodimerisation/polymerization of NVC induced in this way can now be seen to be a special case of the wide range (5) of photoinduced electron transfer reactions of amines, referred to in Section I. Generally, these reactions may be represented as indicated below, although products vary according to the nature of the amine substituents and the sensitiser:

$$RCH_2NR'_2 + [Sens]^* \rightarrow [RCH_2\overset{\cdot+}{N}R'_2][Sens^{\cdot-}]$$
$$\downarrow$$
$$Products \longleftarrow R\dot{C}HNR'_2 + \dot{S}ens{-}H$$

Of course, the detailed reaction pathways leading to firstly, a reactive ion pair and secondly, to free radicals or products will involve competing photophysical processes, exciplex formation, and solvation phenomena, as described earlier, but recent detailed studies, described in the following section, have helped to characterise the potential of the technique for initiation of free radical polymerization.

B. Initiation of Free Radical Polymerization

1. Benzophenone/Triethylamine

Sandner et al. (21) observed that addition of small amounts of triethylamine (0.02 M) greatly enhanced the photo-induced polymerization of methyl acrylate (1.0 M in tert-butanol, nitrogen-flushed) in the presence of benzophenone (0.02 M). Photoinitiation was not effected by triethylamine alone, nor by triphenylamine, isopropanol, or benzhydrol (all 0.02 M) in the presence of benzophenone. Quantum yields measured for benzophenone disappearance indicated that methyl acrylate itself acted as a quencher of photoexcited benzophenone, effectively suppressing hydrogen abstraction from tert-butanol, and benzhydrol. However this quenching action had only a marginal effect on the efficiency of interaction of photoexcited benzophenone with triethylamine, suggesting that something other than direct hydrogen abstraction from the amine was involved. Consistent with this view was the observation that triphenylamine inhibited photoreduction of benzophenone in *tert*-butanol. Efficient formation of an excited electron-transfer complex from triplet benzophenone and ground-state amine was postulated, but initiation by the complex itself, perhaps by an anionic pathway, was discounted in view of the ineffectiveness of triphenylamine as co-initiator. In the case of triethylamine, it was proposed that initiating radicals resulted from collapse of the complex following proton transfer.

$$Ph_2C=O(T_1) + (C_2H_5)_3\ddot{N} \longrightarrow [Ph_2\dot{C}-\bar{O}\ \overset{+}{N}(C_2H_5)_3]$$

$$\downarrow$$

$$Ph_2\dot{C}-OH + CH_3\dot{C}HN(C_2H_5)_2$$

Since analogous proton transfer cannot take place from triphenylamine, the complex formed in this case presumably reverts to ground-state ketone and amine, the overall effect being efficient quenching of triplet benzophenone.

2. Fluorenone-Amine Systems

The kinetics of photoinduced polymerization of methyl methacrylate in the presence of fluorenone (FL) and N,N-dimethyl ethanolamine (DME) have been studied in some detail (*85*). The experiments were carried out at 30° C, using degassed solutions sealed under high vacuum, and employing a 250 W medium-pressure mercury discharge lamp as the source of ultra-violet light. Polymerization rates exhibited half-order dependence on incident light intensity and fluorenone concentration (in the range $0-1.5 \times 10^{-4}$ M), and were first-order with respect to monomer concentration, as determined from initial rates measured for various concentrations of methyl methacrylate in benzene; thus a radical chain mechanism was indicated.

As a function of N,N-dimethyl ethanolamine concentration, the polymerization rate passed through a maximum, as shown in Fig. 2. Such behaviour parallels that observed for photoreduction of fluorenone by other tertiary amines, e.g. the quantum yield for photoreduction by triethylamine has been measured (*5*) as 0.09 in neat amine, 0.9 in

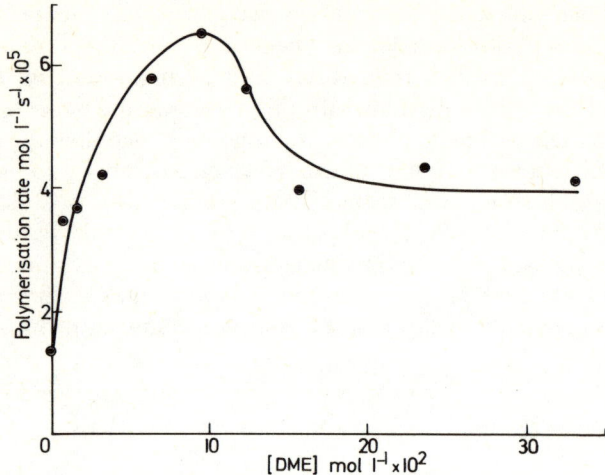

Fig. 2. Dependence of photo-induced polymerization rate on N,N-dimethylethanolamine concentration; [fluorenone] = 1.14×10^{-4} M, [methyl methacrylate] = 3.9 M in benzene

0.1 M amine (in cyclohexane) and 0.6 in 0.02 M amine; the factors which determine the quantum efficiency of these reactions have been fully discussed elsewhere (5). Molecular weight determinations in the FL/DME/MMA system gave a value for $k_p/k_t^{1/2} = 0.054 \,\text{mol}^{-1/2}\, l^{1/2}\, s^{-1/2}$ indicating that transfer and termination processes are not particularly affected by the initiating components or their photoredox products. Almost all tertiary amines, having α-hydrogen atoms, may be used conveniently, and other carbonyl compounds, notably N-methyl acridone (NMA) which has a much higher molar extinction coefficient than fluorenone, are also effective.

The scope of this initiation reaction may be widened by using an arylamino-acetic acid derivative as the electron donor, with the same sensitisers, and of these indol-3yl-acetic acid (IAA) is experimentally convenient. In these cases it is thought that the initiating radicals arise via decarboxylation following the normal photoinduced electron transfer. Quantum yields for initiation by the combination FL/DME, FL/IAA and NMA/IAA lie in the range 0.1—0.2 and it is clear that this technique affords a new versatile range of initiating systems for radical polymerization.

V. References

1. Wells, C. H. J.: Introduction to molecular photochemistry. London: Chapman and Hall 1972.
2. Simons, J. P.: Photochemistry and spectroscopy. London: Wiley-Interscience 1971.
3. Lutz, H., Duval, M. C., Bréhéret, E., Lindqvist, L.: J. Phys. Chem. **76**, 821 (1972).
4. Previtali, C. M., Scaiano, J. C.: J. Chem. Soc. Perkin II, 1667 (1972).
5. Cohen, S. G., Parola, A., Parsons, G. H.: Chem. Rev. **73**, 141 (1973).
6. Ottolenghi, M.: Accounts Chem. Res. **6**, 153 (1973).
7. Heine, H.-G., Rosenkranz, H.-J., Rudolph, H.: Angew. Chem. Int. Edit. **11**, 974 (1972).
8. Oster, G., Yang, N.-L.: Chem. Rev. **68**, 125 (1968).
9. Rabek, J. F.: Photochem. Photobiol. **7**, 5 (1968).
10. Kosar, J.: Light-sensitive systems, Chapter 5. New York: Wiley 1965.
11. Bevington, J. C.: Radical polymerization, p. 77. London: Academic Press 1961.
11a. Mochel, W. E., Crandall, J. L., Peterson, J. N.: J. Am. Chem. Soc. **77**, 494 (1955).
12. Chen, C.-T., Huang, W.-D.: J. Chin. Chem. Soc. (Taipei), **16**, 46 (1969).
13. Heine, H.-G., Fuhr, K., Rudolph, H., Schnell, H.: S. Afr. Pat. 6904, 724 (Chem. Abstr. 1970, **73**, 46220).
14. Sun Chemical Corp. Br. Pat. 1 198 259 (Chem. Abstr. 1970, 73, 57311).
15. Plambeck, L.: U.S. Pat. 2760863 (Chem. Abstr. 1956, **50**, 16498).
16. Chinmayanandam, R. B., Melville, H. W.: Trans. Faraday Soc. **50**, 73 (1954).
17. Closs, G. L., Paulson, D. R.: J. Am. Chem. Soc. **92**, 7229 (1970).
18. Ledwith, A., Russell, P. J., Sutcliffe, L. H.: J. Chem. Soc. Perkin II, 1925 (1972).
19. Heine, H.-G.: Tetrahedron Letters 4755 (1972).
20. Hutchison, J., Ledwith, A.: Polymer **14**, 405 (1973).
21. Sandner, M. R., Osborn, C. L., Trecker, D. J.: J. Polymer. Sci. A–1, **10**, 3173 (1972).
22. Kurien, K. C.: J. Chem. Soc. B, 208 (1971).
23. Baum, E. J., Hess, L. D., Wyatt, J. R., Pitts, J. N., Jr.: J. Am. Chem. Soc. **91**, 2461 (1969).
24. Evans, D. F.: J. Chem. Soc. 1351 (1957).
25. Solly, R. K., Benson, S. W.: J. Am. Chem. Soc. **93**, 1592 (1971).
26. Braun, D., Becker, K. H.: Angew. Makromol. Chem. **6**, 186 (1969); — Makromol. Chem. **147**, 91 (1971).
27. Sheehan, J. C., Wilson, R. M.: J. Am. Chem. Soc. **86**, 5277 (1964).
28. Petropoulos, C. C.: J. Polymer Sci. A–2, 69 (1964).
29. Schönberg, A., Fateen, A. K., Omran, S. M. A. R.: J. Am. Chem. Soc. **78**, 1224 (1956).
30. Tsunooka, M., Kusube, M., Tanaka, M., Murata, N.: Kogyo Kagaku Zasshi **72**, 287 (1969).
31. Tsuda, K., Otsu, T.: Bull. Chem. Soc., Jap. **39**, 2206 (1966).
32. — Kose, K.: Makromol. Chem. **161**, 267 (1972).
33. Walling, C.: Free radicals in solution, p. 239. New York: Wiley 1957.

34. Monroe, B. M.: Adv. Photochemistry **8**, 77 (1970).
35. Turro, N. J.: Molecular photochemistry. New York: Benjamin 1965.
36. Neckars, D. C.: Mechanistic organic photochemistry. New York: Reinhold 1967.
37. Porter, G., Suppan, P.: I. U. P. A. C. Symposium on Organic Photochemistry, p. 499. Strassbourg 1964.
38. Gibian, M. J.: Tetrahedron Letters 5331 (1967).
39. Filipescu, N., Minn, F. L.: J. Am. Chem. Soc. **90**, 1544 (1968).
40. Weiner, S. A.: J. Am. Chem. Soc. **93**, 425 (1971).
41. Pitts, J. N., Letsinger, R. L., Taylor, R. P., Patterson, J. M., Recktenwald, G., Martin, R. B.: J. Am. Chem. Soc. **81**, 1068 (1959).
42. Wilkinson, F.: J. Phys. Chem. **66**, 2569 (1962).
43. Tickle, K., Wilkinson, F.: Trans. Faraday Soc. **61**, 1981 (1965).
44. Neely, W. C., Dearman, H. H.: J. Chem. Phys. **44**, 1302 (1966).
45. Oster, G., Oster, G. K., Moroson, H.: J. Polym. Sci. **34**, 671 (1959).
46. Block, H., Ledwith, A., Taylor, A. R.: Polymer **12**, 271 (1971).
47. Hammond, G. S., Hamilton, E. J., Weiner, S. A., Hefter, H. J., Gupta, A.: I. U. P. A. C. XXIIIrd Congress, Boston 1971, Vol. 4, p. 257. London: Butterworth 1971.
48. North, A. M.: The kinetics of free radical polymerization, p. 79. Oxford: Pergamon Press 1966.
49. Elad, D.: Fortschr. Chem. Forsch. **7**, 528 (1966).
50. Jacobs, R. L., Ecke, G. G.: J. Org. Chem. **28**, 3036 (1963).
51. Elad, D., Youssefyeh, R. D.: J. Org. Chem. **29**, 2031 (1964).
52. Hutchison, J., Lambert, M. C., Ledwith, A.: Polymer **14**, 250 (1973).
53. Breslow, R.: Chem. Soc. Rev. **1**, 553 (1972).
54. — Baldwin, S., Flechtner, T., Kalicky, P., Liu, S., Washburn, W.: J. Am. Chem. Soc. **95**, 3251 (1973).
55. Hutchison, J., Ledwith, A.: Unpublished results.
56. Cooper, W., Vaughn, G., Miller, S., Fielden, M.: J. Polymer. Sci. **34**, 651 (1959).
57. Hallensleben, M. L.: Europ. Polymer J. **9**, 227 (1973).
58. Gaylord, N. G., Dixit, S. S.: J. Polymer Sci. B **9**, 823 (1971).
59. Bunbury, D. L., Wang. C. T.: Can. J. Chem. **46**, 1473 (1968).
60. — Chuang, T. T.: Can. J. Chem. **47**, 2045 (1969); but see also, Inoe, H., Takido, S., Somemiya, T., Nomura, Y.: Tet. Letters, 2755 (1973).
61. Wells, C. F.: Discuss. Faraday Soc. **29**, 219 (1960).
62. — Trans. Faraday Soc. **57**, 1719 (1961).
63. — Trans. Faraday Soc. **57**, 1703 (1961).
64. Eigenmann, G.: Helv. Chim. Acta, **46**, 804 (1963).
65. Cooper, H. R.: Trans. Faraday Soc. **62**, 2865 (1966).
66. Atkinson, B., Di, M.: Trans. Faraday Soc. **54**, 1331 (1958).
67. Carapelluci, P. A., Wolf, H. P., Weiss, K.: J. Am. Chem. Soc. **91**, 6335 (1969); Creed, D., Hales, B. J., Porter, G.: Proc. Roy. Soc. Lond. A **334**, 505 (1973).
68. Schulte-Frohlinde, D., Sonntag, H.: Z. Physik. Chem. (Frankfurt) **44**, 314 (1965).
69. Bolland, J. L., Cooper, H. R.: Proc. Roy. Soc. (A), **225**, 405 (1954).
70. Walker, P.: J. Chem. Soc. 5545 (1963).
71. Wells, C. F.: J. Chem. Soc. 3100 (1962).
72. Kalle and Co. A-G.: Ger. Pat. 1, 046, 317 (1958); — Chem. Abstr. **55**, 1074 (1961).
73. E. I. du Pont de Nemours and Co., U.S. Pat. 2,951,758 (1960); — Chem. Abstr. **55**, 1946 (1961).

74. Dunlop Rubber Co. Ltd.: Brit. Pat. 861,438 (1961); — Chem. Abstr. **55**, 17100 (1961).
75. E. I. du Pont de Nemours and Co., U.S. Pat. 3,046,127 (1962); — Chem. Abstr. **57**, 15374 (1962).
76. E. I. du Pont de Nemours and Co., U.S. Pat. 3,024,180 (1962); — Chem. Abstr. **56**, 14446 (1962).
77. Anwaraddin, Q., Santappa, M.: J. Polymer Sci. B **5**, 361 (1967).
78. Ledwith, A., Ndaalio, G., Taylor, A. R.: Manuscript in preparation.
79. Rubin, M. B., Zwitkowits, P.: J. Org. Chem. **29**, 2362 (1964); — Tetrahedron Letters 2453 (1965).
80. Tichy, J. R.: J. Polymer Sci. **33**, 353 (1958).
81. Ledwith, A.: Accounts. Chem. Res. **5**, 133 (1972).
82. Hyde, P., Ledwith, A.: Molecular complexes, Vol. 2, R. Foster. (Ed.). London: Logos Press (in press).
83. Lambert, M. C.: Ph. D. Thesis, Univ. of Liverpool (1973).
84. Tada, K., Shirota, Y., Mikawa, H.: Macromolecules **6**, 9 (1973).
85. Ledwith, A., Purbrick, M. D.: Polymer **14**, 521 (1973).

Received July 23, 1973

Chemical Transformations of Cellulose

L. S. GAL'BRAIKH and Z. A. ROGOVIN

Moscow Textile Institute, Department of Chemical Fibre Technology,
Moscow B-419, USSR

Tabe of Contents

I. Introduction: Synthesis of Cellulose-based Mixed Polysaccharides: A Study of Some of their Properties 88

II. Synthesis of Mixed Polysaccharides Containing Repeating Units of Amino Sugars and their N-Substituted Derivatives 102

III. Synthesis of Mixed Polysaccharides: C-Alkyl Derivatives of Deoxycellulose 104

IV. Synthesis of Unsaturated Cellulose Derivatives and their Chemical Transformations 106

V. Synthesis of Organoelement Derivatives of Cellulose 114
   1. Synthesis of Silicon-Containing Cellulose Esters 114
   2. Synthesis of Fluorine-Containing Cellulose Ethers and Esters . . . 115
   3. Synthesis of New Types of Phosphorus-Containing Cellulose Esters . 117
   4. Synthesis of Tin-Containing Cellulose Esters 123

VI. Discussion of the Principles Governing the Synthesis of Cellulose Esters by the Trans-esterification Reaction 124

VII. References 128

I. Introduction:
Synthesis of Cellulose-based Mixed Polysaccharides: A Study of Some of their Properties

Cellulose is one of the most important and most abundant natural polymers. This circumstance and certain specific features of its molecular and supermolecular structures (the linear, strictly regular structure of the macromolecules, the great rigidity of the polymeric chains, the possibility of formation elements of the supermolecular structure having a high degree of ordering, i.e. crystallites, etc.) are the factors that have stimulated extensive work on the chemistry, physicochemistry and technology of cellulose and its derivatives.

Scientists in the department of chemical-fiber technology at the Moscow Textile Institute have for a number of years been conducting systematic studies of the synthesis and preparation of new types of cellulose derivatives by applying new methods of chemical modification (mainly graft polymerization). They are seeking new cellulosic materials possessing potentially useful properties, such as fire resistance, or suitable as ion-exchange, bactericidal, hemostatic and other agents.

The present work summarizes the some results of these studies. The article describes some new types of cellulose ethers and esters, a number of mixed polysaccharides based on cellulose, which have been synthesized by the authors and their coworkers, and also discusses the principles governing a number of reactions for effecting the chemical transformation of cellulose, i.e. nucleophilic substitution, *trans*-esterification, ionic and free-radical addition, which enable new types of cellulose derivatives to be synthesized.

A separate and promising trent in the chemical modification of cellulose is the synthesis of graft copolymers of cellulose with various vinyl and diene monomers. This is a big problem of independent interest; our results in this line obtained in recent years have been published in special revies articles (3).

Here we discuss new types of cellulose derivatives synthesized by nucleophilic substitution of mixed polysaccharides containing repeating units of 2,3- and 3,6-anhydro sugars, units of amino sugars, their N-alkyl(aryl)- and N-carboxylalkyl(aryl) derivatives, alkyl derivatives and phenylbarenyl derivatives of deoxyglucose, elimination reactions of unsaturated derivatives containing multiple C–C bonds in the pyranose ring, and in the alkyl (acyl) radical of cellulose ethers and esters, and also new types of cellulose derivatives containing the elements silicon, fluorine, phosphorus, and tin.

In the general study of the modification of the properties of cellulose through chemical transformations of hydroxyl groups special importance is attached to the effects of fine differences in the composition and structure of the individual repeating units of the macromolecule on the physical and chemical properties of cellulose. A transformation must be accomplished from cellulose, a homopolysaccharide made up of repeating units of glucose, to heteropolysaccharides, mixed polysaccharides containing in addition to glycose units residues of other sugars differing from glucose in the configuration of the functional groups at the secondary carbon atoms of the repeating unit.

A reaction of great theoretical and practical interest is nucleophilic substitution; this reaction can be used to prepare new cellulose derivatives of different classes, e.g. ethers and esters, deoxy and anhydro derivatives. For the synthesis of cellulose derivatives by means of nucleophilic substitution, the starting materials must be compounds in which the bond between the carbon atom of the repeating unit of the macromolecule and the substituent is strongly polarized. Such compounds include certain cellulose esters (predominantly, those of strong acids: aryl and alkyl sulfonic acids, nitric and sulfuric acids, and also halodeoxy derivatives of cellulose). The change in the nature of the group being displaced, its position in the repeating unit, and the nature of the nucleophilic reagent and solvent can obviously influence the direction of the reaction and its stoichiometry.

We give below the basic data obtained from a systematic study of the factors determining the mechanism and stoichiometry of the nucleophilic substitution reactions used for the synthesis of cellulose derivatives.

These nucleophilic substitution reactions can in theory proceed intra- or intermolecularly (4). The structure of the repeating unit of the cellulose macromolecule is such that in an intramolecular reaction, two types of anhydro derivatives may form: mixed polysaccharides, whose repeating units contain α-oxide rings (2,3-anhydro rings), and anhydro rings of the tetrahydrofuran type (3,6-anhydro rings).

Synthesis and Properties of 2,3- and 3,6-Anhydro Derivatives of Cellulose

One of the most popular methods of synthesizing anhydro derivatives of monosaccharides containing α-oxide rings is the alkaline saponification of sugar tosylates (5). We have studied the possibility of using this method for synthesis of the 2,3-anhydro derivative of cellulose. The starting compound used for this purpose is 2(3)-O-tosyl-6-O-trityl-cellulose (I), and its alkaline saponification must proceed according to the scheme

$$\text{(I)} \xrightarrow{\text{CH}_3\text{ONa(CH}_3\text{OH)} \atop \text{KOH(CH}_3\text{OH)}} \quad (1)$$

$$\text{Tr} = (\text{C}_6\text{H}_5)_3\text{C}; \quad \text{Ts} = \text{CH}_3\text{C}_6\text{H}_4\text{SO}_2$$

A study of the composition of the reaction products has shown that, apart from the reaction leading to the formation of 2,3-anhydro rings, there occur elimination reactions of tosyloxy groups, as a result of which the reaction product contains double bonds and carbonyl groups (6, 7). About 80% of the α-oxide rings formed are simultaneously opened. Thus, the alkaline saponification of (I), in contrast to the analogous treatment of monosaccharide tosylates which leads to the quantitative formation of an anhydro derivative (8), does not enable one to prepare a mixed polysaccharide containing a considerable amount of 2,3-anhydro rings.

This is probably because of the hindered transition of the repeating unit from the conformation C1, in which it is present in the cellulose macromolecule and where conditions do not favor the occurrence of a nucleophilic substitution reaction (the trans-diaxial arrangement of tosyloxy and hydroxyl groups at C_2 and C_3), into a conformation that would satisfy these requirements. These difficulties are associated with the specific structural features of cellulose, and they are made worse by the presence of a bulky trityl group at C_6 in compound (I). Steric hindrances can probably be eliminated by using as the starting material a compound containing no trityl groups, e.g. 2(3)-O-tosylcellulose (II) (9, 10). The interaction of (II) with lithium acetate in dimethylsulphoxide (DMSO) results in the formation of a partially acetylated mixed polysaccharide where up to 60% of the repeating units contain α-oxide rings (Table 1).

The data obtained also show that in going from sodium methylate and potassium hydroxide to lithium acetate, a reagent of relatively low nucleophilicity and comparatively high basicity, the intensity of side reactions of elimination is sharply reduced.

Table 1. Composition of the products of interaction of 2(3)-O-tosylcellulose (II) with a solution of lithium acetate in DMSO

Composition II	Composition of reaction product					
Sulphur content (%)	DS[a] Sulphur content (%)	DS of tosyloxy groups	Content of bound CH$_3$COOH (%)	DS of acetyl groups	Number of α-oxide rings per repeating unit	Number of double bonds per repeating unit
8.30	0.70 1.50	0.08	5.60	0.17	0.49	0.01
8.60	0.74 1.70	0.09	5.97	0.19	0.48	0.03
8.19	0.68 0.80	0.04	2.13	0.06	0.60	0.0

[a] DS is the degree of substitution (the number of substituents per repeating unit).

From a study of 2,3-anhydro derivative of cellulose by infrared spectroscopy combined with the determination of the monosaccharide composition of the products of its acetylation and interaction with sodium methylate, we conclude that the mixed polysaccharide synthesized contains glucose and 2,3-anhydromannose units.

The chemical transformations of a mixed polysaccharide containing 2,3-anhydromannose units are based on the opening of the α-oxide ring under the action of certain nucleophilic reagents. These transformations are of interest not only as a method of identifying its monosaccharide composition but also as a way to obtain new types of mixed polysaccharides whose structure is close to that of such natural polysaccharides as cellulose and chitosan, but differs from them in the configuration of the substituents at the secondary carbon atoms of the repeating unit a mixed polysaccharide containing altrose and glucose units (III) or in the configuration and position of the substituents a mixed polysaccharide containing units of glucose and 3-deosy-3-aminoaltrose (IV):

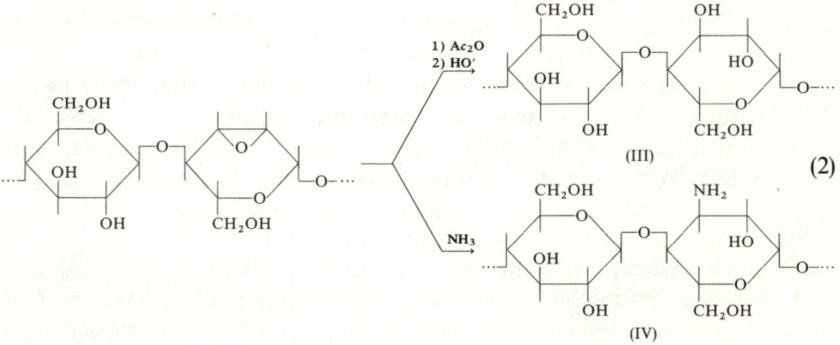

(2)

Table 2. Composition of the products of acetylation of 2,3-anhydro derivatives of cellulose

Number of anhydro rings per repeating unit	Content of bound CH_3COOH in acetylation product (%)	Content of indicated compounds in mixed polysaccharide (mol.-%)	
		altrose	glucose
12	62.0	10	90
26	58.0	25	75
45	57.6	42	58

Data on the monosaccharide composition of the products of acetylation of 2,3-anhydro derivatives of cellulose are given in Table 2.

Another type of cellulose anhydro derivative, a mixed polysaccharide containing glucose and 3,6-anhydroglucose units, has been synthesized by the alkaline saponification of cellulose tosylate (V) (*11, 12*) and 6-O-tosyl-2,3-di-O-acetylcellulose (VI) (*13*):

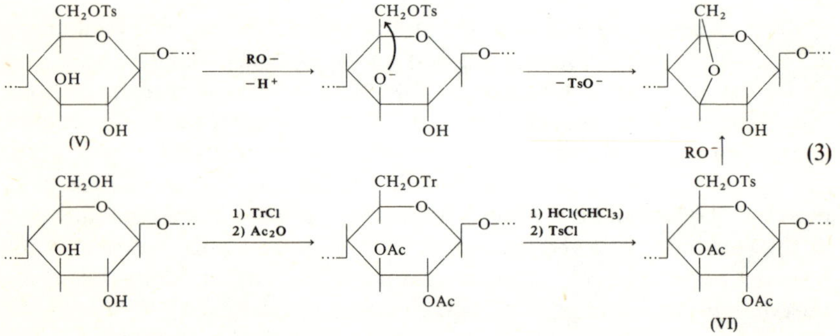

$$ (3) $$

The monosaccharide composition of the products of the alkaline saponification of cellulose tosylates containing tosyloxy groups at C_6 has been established on the basis of data obtained from a study of the composition of the products of acetylation, nitration, and periodate oxidation, and from quantitative paper chromatography of the products of complete hydrolysis of the mixed polysaccharides synthesized. The content of 3,6-anhydroglucose units in the 3,6-anhydro derivatives of cellulose was found to be 50–70 mol.-% (*10–13*).

Studies of cellulose anhydro derivatives containing 2,3- and 3,6-anhydro rings by means of infrared spectroscopy (*14, 15*) have made it possible to draw inferences about the nature of the conformational changes that occur in the repeating units during the formation of anhydro rings.

The spectra of mixed polysaccharides containing 2,3-anhydro-mannose units show in the 700–950 cm^{-1} region a new band at 855 cm^{-1} characteristic of C_1–$H_{(e)}$ bonds. The change of the spatial orientation of the C_1–H bond from axial to equatorial has evidently occurred as a result of the conversion of the repeating unit from conformation C1 to a half-chair conformation:

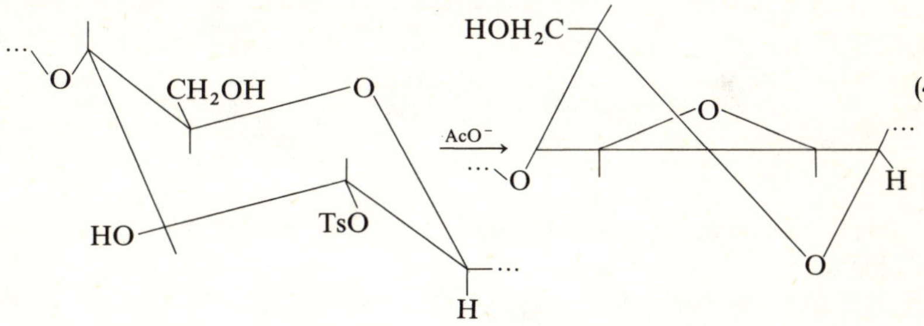

(4)

In the spectra of mixed polysaccharides containing 3,6-anhydro-glucose units there appear new absorption bands at 920, 840 and 800 cm^{-1}. The band at 800 cm^{-1} must be attributed, according to Barker and Stephens (*16*), to the vibrations of the tetrahydrofuran (3,6-anhydro) ring. The band at 840 cm^{-1} is found in the spectra of glycopyranosides whose molecule has an equatorial hydrogen atom at C_1. The appearance of this band in the spectrum of a mixed polysaccharide may be explained by the fact that during the formation of a 3,6-anhydro ring the repeating unit changes from conformation C1 to conformation 1C:

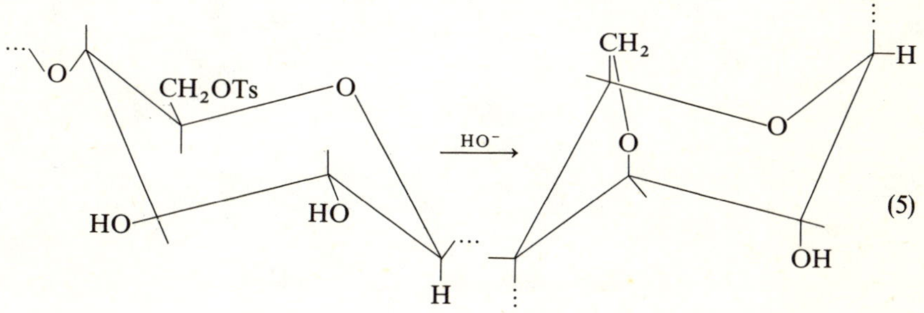

(5)

Interesting data have been obtained from the determination of the content of methoxyl groups in the products of the saponification of cellulose tosylate with a methanol solution of KOH (Table 3).

Table 3. Content of methoxyl groups in 3,6-anhydro derivatives of cellulose

Amount of tosyloxy groups in starting cellulose tosylate, mole/repeating unit		Content of CH_3O groups in mixed polysaccharide (%)		Content of CH_3O groups (%)	
At C_6	At $C_{2,3}$	Calculated	Found	After treatment with 0.1 N H_2SO_4 for 4.5 h	In products of complete hydrolysis of mixed polysaccharide
0.800	0.292	5.76	10.75	5.98	5.15
0.754	0.246	4.88	10.60	5.64	4.34
0.723	0.227	4.50	10.38	4.71	3.92

Methoxyl groups stable to acid treatment occur in a mixed polysaccharide as a result of the opening by a methoxy ion of the 2,3-anhydro rings formed by the splitting-off of the tosyloxy groups situated at the secondary carbon atoms of the repeating unit in a similar way to Scheme 2. The presence in a mixed polysaccharide of methoxyl groups unstable to acid treatment is probably accounted for as follows. According to literature data (17, 18), a bicyclic system containing 3,6-anhydro and 1,5-pyranose rings is strained and as a result under certain conditions, 3,6-anhydroglyco-pyranosides are converted to the corresponding furanosides or to an a l-form. When a 3,6-anhydro ring is forming in the polymeric chain, this strain can be increased by steric hindrances so as to induce the conformational rearrangement of the repeating unit. This may cause some of the 3,6-anhydroglucose units to change to an opened form:

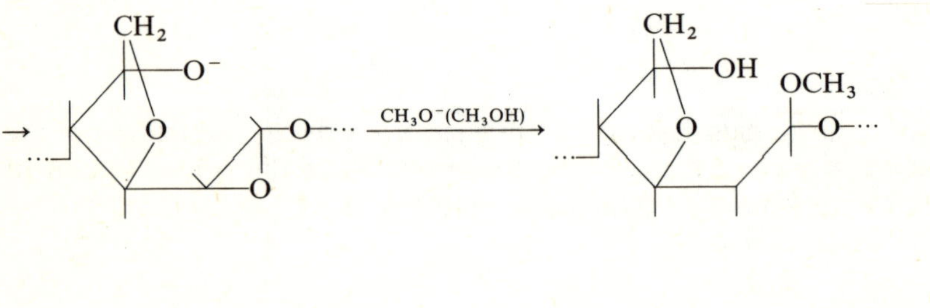

(6)

This involves the appearance at C_1 of methoxyl groups, which in contrast to ethers, are cleaved off on acid treatment.

The chemical transformations that lead to the conversion of cellulose to mixed polysaccharides differing from cellulose in the conformation of the pyranose ring and the number and configuration of the hydroxyl groups of the repeating unit of the macromolecule, may exert a considerable effect on the structure of the material as well as on its important chemical properties (rate of acetylation and O-alkylation of OH groups, stability of the acetal linkage) and physicochemical indices (solubility of modified preparations of cellulose and cellulose ethers and esters).

We discuss below data obtained from an investigation of certain chemical and physicochemical properties of mixed polysaccharides containing altrose and glucose units (III) and units of 3,6-anhydroglucose and glucose (IV), and of their derivatives.

The effect of the spatial arrangement of the hydroxyl groups of the repeating unit on reactivity has been investigated in comparative studies of the acetylation and O-nitration of cellulose and mixed polysaccharides (III) and (IV) (*19, 20*). The data obtained are presented in Tables 4 and 5.

Table 4. Effect of the number of altropyranose units in mixed polysaccharide (III) on the rate constant of the acetylation reaction[a]

Content of altrose units (mol.-%)	0 (Cellulose)	10	25	31	42
$K \cdot 10^2$ min^{-1}	1.632	0.567	0.367	0.317	0.184

[a] The conditions of acetylation are as follows: modulus 40; concentration of H_2SO_4: 0.125 g/litre; temperature: 30° C.

Table 5. Effect of the number of 3,6-anhydroglucose units in mixed polysaccharide (IV) on the rate constant of the acetylation reaction[a]

Content of 3,6-anhydroglucose units	$K \cdot 10^{-2}$ min^{-1}
0.0 (Cellulose)	6.40
17.0	4.90
70.0	4.30

[a] The conditions of acetylation are as follows: activation by CH_3COOH vapor.; modulus 25; concentration of H_2SO_4: 2 g/litre; temperature: 40° C.

According to these data, an increase in the content of altropyranose units in the macromolecule of mixed polysaccharide (III) leads to a sharp fall in its acetylation rate as compared with cellulose. Similar results have been obtained for the nitration reaction of (III). An analogous relationship is observed for mixed polysaccharide (IV) upon acetylation preceded by the activation of the starting material.

The differences found in the acetylation rates of cellulose and mixed polysaccharide (III), which differs from cellulose only in the configuration of the substituents (OH groups) at C_2 and C_3, can be explained on the basis of the principles of conformational analysis. If, in the synthesis of mixed polysaccharide (III) accomplished by the opening of the α-oxide rings in the 2,3-anhydro derivative of cellulose, the altropyranose unit assumes conformation C1, the OH groups at C_2 and C_3 must be in the *trans*-diaxial position (i.e. the change in the configuration of the substituents at the secondary C atoms simultaneously changes their spatial orientation with respect to the ring). According to Reeves (*21*), altropyranose is characterized by the presence of equilibrium between two conformations (C1⇌1C). When the repeating unit is converted to conformation 1C, the hydroxy-methyl group will assume the axial position.

Thus, in both possible conformations the altropyranose units contain axial hydroxyl groups. The presence of these groups is confirmed by the appearance in their spectra of the absorption band of C–H_e 820 cm^{-1}.

The presence of axial OH groups causes a reduction in the rate of esterification of mixed polysaccharide (III).

For mixed polysaccharide (IV), which contains glucose and 3,6-anhydroglucose units, the difference in the rate of acetylation as compared with cellulose is probably due to several factors.

The reduction in the total number of hydroxyl groups and the disturbance of the regular structure of a mixed polysaccharide results in a less intensive intermolecular interaction. There is thus a certain increase in the rate of the process, manifested in the fact that the acetylation reaction takes place without preliminary activation. At the same time the formation of 3,6-anhydro rings reduces the number of primary hydroxyl groups, which are most reactive in esterification reactions, with a corresponding decrease in the rate of acetylation. This effect is clearly observed in the case of polysaccharide (IV) containing 70 mol.-% of 3,6-anhydroglucose units. The slowing of the acetylation rate may also be associated with the axial arrangement of the OH groups at C_2 in the 3,6-anhydroglucose units; they are probably in conformation 1C or a skew conformation close to it.

In contrast to esterification reactions, where the reactivity of the hydroxyl groups of polysaccharides is primarily governed by their steric accessibility, the reactivity of the hydroxyl groups of mono- and polysaccharides in reactions of O-alkylation is controlled mainly by the mobility of the hydrogen atoms of the OH groups, and this is determined by the magnitude of the dissociation constant.

We have studied the relative reactivity of cellulose and mixed polysaccharides (III) and (IV) in reactions with aqueous and alcoholic solutions of NaOH. The data on the composition of alkali compounds of polysaccharides are given in Table 6. As seen from the data presented in Table 6, the amount of bound alkali in preparations of alkaline compounds of mixed polysaccharides is less than in alkali cellulose obtained under the same conditions, the DS with respect to NaOH decreasing with increasing content of altrose units in polysaccharide (III) and of 3,6-anhydroglucose units in polysaccharide (IV).

According to literature data (22), the highest acidity (dissociation constant) in α- and β-methylglucosides is exhibited by the OH group at C_2, the dissociation constant for β-methylglucoside being higher than

Table 6. Effect of number of altrose or 3,6-anhydroglucose units in the macromolecule of a mixed polysaccharide on the composition of alkaline compounds of polysaccharides (III) and (IV)

Type of polysaccharide	Amount of altrose units (mol.-%)	Amount of 3,6-anhydroglucose units (mol.-%)	Characteristics of alkali compound of polysaccharide	
			DS	Amount of reacted hydroxyl groups (%)
Cellulose	0	0	0.85	28.4
(III)	10.0	0	0.65	21.6
(III)	31.0	0	0.15	5.0
Cellulose	0	0	1.03	34.4
(IV)	0	19.0	0.77	29.4
(IV)	0	42.5	0.59	27.4

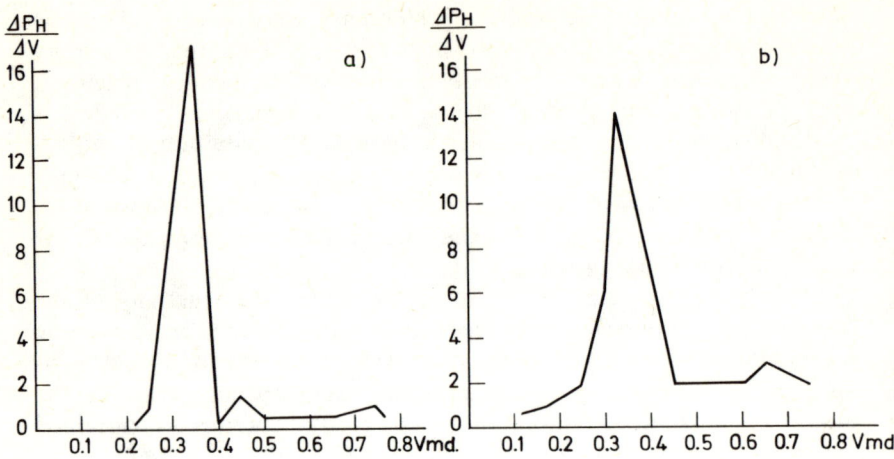

Fig. 1. Differential potentiometric titration of (a) α-methylglucoside and (b) α-methylaltroside

for the α-anomer, due to the increase in the probability that the OH group at C_2 will form a hydrogen bond with the axially arranged oxygen atom of the glucoside hydroxyl.

The dissociation constant of the OH group at C_2 in the repeating unit of the macromolecule of polysaccharides may also be reduced as a result of the formation of a hydrogen bond. Thus, according to the available evidence (22), the acidity of the OH group at C_2 of the repeating unit of amylose, which reacts with alkali in conformation B1, is lower than in cellulose. The data suggest that the acidity of the OH groups at C_2 in the altropyranose units of 3,6-anhydroglucose is lower than the acidity of such groups in glucopyranose units due to incorporation of the axially arranged OH groups at C_2 into a hydrogen bond with an equatorial glucoside oxygen atom in polysaccharide (III), or with the oxygen atom of the pyranose ring in (IV).

This assumption has been experimentally verified by determining the relative acidity of the OH groups of the model compounds methyl-α-D-gluco- and altropyranoside by the method of nonaqueous titration.

The differential titration curves are shown in Fig. 1. As reported (29), the first peak on each curve reflects the neutralization of the OH group at C_2. A comparison of the peaks shows that the degree of dissociation of the OH group at C_2 is lower in α-methyl-D-altroside than in α-methyl-D-glucoside.

To elucidate the effect of the change in the configuration of substituents in the repeating unit and also of the structure of the repeating

Table 7. Rate constants of hydrolysis of cellulose and mixed polysaccharides (III) and (IV)

Type of polysaccharide	Amount of altrose units (mol.-%)	Amount of 3,6-anhydro glucose units (mol.-%)	Reaction temperature (°C)	$K \cdot 10^2$, hr^{-1}	Ratio of constants K_p/K_c
Cellulose	0	0	40	24.0	1.00
(III)	10.0	0	40	31.0	1.29
(III)	31.0	0	40	48.0	2.00
Cellulose	0	0	20	12.88	1.00
(IV)	0	8.3	20	16.40	1.27
(IV)	0	42.6	20	26.07	2.01

unit and its conformation on the stability of the acetal linkage, we carried out a study into the process of hydrolysis in a homogeneous medium of cellulose and mixed polysaccharides (III) and (IV) (23, 24). The data obtained are presented in Table 7. As these data show, the presence of altrose and 3,6-anhydroglucose units in the macromolecules of (III) and (IV) leads to an increase in the rate of hydrolysis of these polysaccharides as compared with cellulose.

On hydrolysis of mixed polysaccharide (III), the axial arrangement of the OH groups at C_2 and C_3 in the altropyranose units in conformation C1, or of an hydroxymethyl group in conformation 1C, changes to equatorial at the stage when a glycosyl cation is formed, which determines the overall rate of hydrolysis. This change weakens the interaction between OH groups and upon rupture of the bond between the altropyranose unit and the neighboring unit a glycosyl cation forms more readily than in the analogous process for cellulose.

The principal factor responsible for the increased rate of hydrolysis of polysaccharide (IV) as compared with cellulose is probably the presence in the macromolecule of hemyacetal linkages between the 3,6-anhydroglucose units, which are in an open form, and the neighboring repeating units:

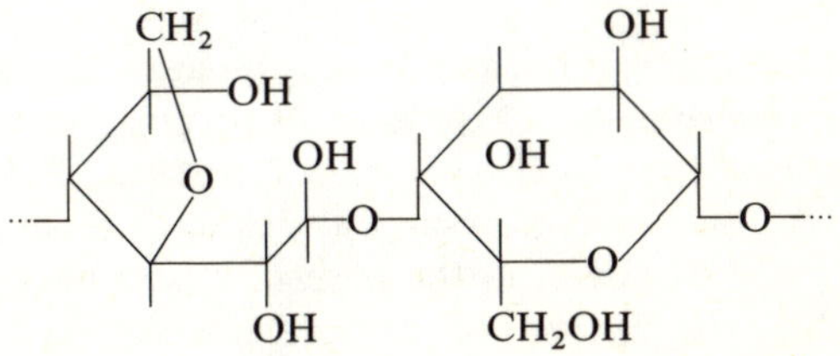

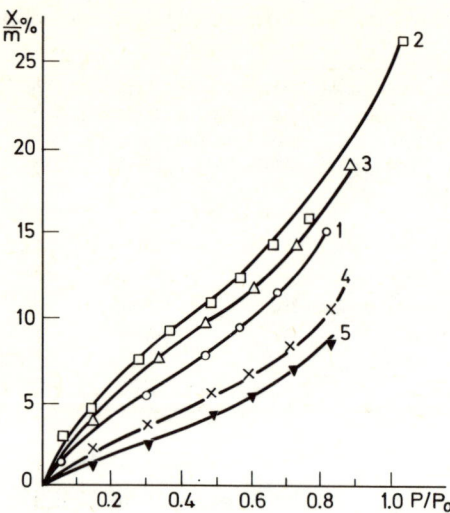

Fig. 2. Isotherms of the sorption of water vapor by the following preparations: *1* hydrate cellulose; *2* mixed polysaccharide (III) containing 10 mol.-% of altropyranose units; *3* mixed polysaccharide (III) containing 31 mol.-% of altropyranose units; *4* cotton cellulose; *5* mixed polysaccharide IV containing 3,6-anhydroglucose units

This assumption would also account for the difference in the ratio of the rates of hydrolysis of cellulose and polysaccharide (IV) as compared with the ratio of the rates of hydrolysis of glycosides and 3,6-anhydroglycosides [the rate of hydrolysis of 3,6-anhydroglycosides is lower than that of the corresponding glycosides (*18, 25*)].

The transformation from cellulose to mixed polysaccharides also induces a change in the structure of the preparation. The data obtained from the radiographic and infrared-spectroscopic investigation of preparations of mixed polysaccharides (III) and (IV) show that even a small number of altrose or 3,6-anhydroglucose units in the macromolecule markedly reduces the intensity of the intermolecular interaction and the degree of ordering of the preparation as compared with the starting cellulose.

To study the structural features of preparations of mixed polysaccharides, we investigated the process of sorption of water vapor on a vacuum sorption setup of the McBain type (Fig. 2). The data show that preparations of mixed polysaccharides sorb a larger amount of moisture than cellulose, even in the case of mixed polysaccharide (IV) where the number of hydroxyl groups decreases as a result of the formation of 3,6-anhydro rings. As the number of 3,6-anhydroglucose units in the macromolecule of mixed polysaccharide (IV) increases, the amount of sorbed moisture also rises; this points to increasing looseness of structure and decreasing intensity of intermolecular interaction. Less moisture is absorbed by the mixed polysaccharide (III) with higher number of altrose units in its macromolecule; this is probably due to the formation of a new system of intra- and intermolecular hydrogen bonds.

Additional data on the character of these structural changes were obtained from an investigation of the sorption of iodine by preparations of mixed polysaccharide (III), and the determination of the heats of swelling in water (Table 8) (26).

Table 8. Sorption of iodine and heats of swelling of cellulose and mixed polysaccharide (III) in water

Type of polysaccharide	Amount of altrose units (mol.-%)	Amount of sorbed iodine (mg/g)	Heat of swelling in water (kcal/mole)
Cellulose	0	50.2	1.95
(III)	10	160.0	2.75
(III)	31	90.0	2.26

We have also studied the solubility of mixed polysaccharides (III) and (IV) in cellulose solvents such as copper-ammoniacal solution and cadoxene; for polysaccharide (IV) we also used a sodium-iron-tartaric complex. The data obtained are given in Table 9. The dissolution of the polysaccharides in these solvents is associated with the formation of complexes between the OH groups of glycol groupings and the metal ions (or complex ions) (27). It follows that with polysaccharide (IV) a sharp fall in solubility will occur because complexes cannot form with the solvent due to the absence of glycol groupings in 3,6-anhydroglucose. With mixed polysaccharide (III) the decrease in solubility is directly connected with configurational changes in the altropyranose units. It is known that, for a complex to form with a Cu^{2+} ion, the distance between the oxygen atoms of the α-glycol grouping must not exceed 3 Å (21). With altropyranose units, the distance between the oxygen atoms of the OH groups at C_2 and C_3 for both possible conformations does exceed this value (it is 3.71 Å for conformation C1, and 3.45 Å for conformation 1C), which excludes the possibility of complex formation.

Table 9. Effect on solubility of the composition of mixed polysaccharides

Type of polysaccharide	Amount of altrose units in (III) or of 3,6-anhydroglucose units in (IV) (mol.-%)	Solvent	Insoluble fraction of starting polysaccharide (% wt.)	Altrose units (3,6-anhydroglucose units) (mol.-%)
Cellulose	0	c/a solution	0.0	0.0
Cellulose	0	Cadoxene	0.0	0.0
Cellulose	0	SITC	0.0	0.0
Mixed polysaccharide (III)	25	c/a solution	53.2	36.0
	31	c/a solution	54.1	45.0
	42	c/a solution	70.4	51.0
Mixed polysaccharide (III)	31	Cadoxene	62.1	Not determined
	42	Cadoxene	75.4	Not determined
Mixed polysaccharide (IV)	12	c/a solution	91.6	12.3
	62	c/a solution	91.2	Not determined
Mixed polysaccharide (IV)	12	Cadoxene	86.9	Not determined
	62	Cadoxene	100.0	Not determined
Mixed polysaccharide (IV)	12	SITC	77.0	Not determined
	62	SITC	100.0	Not determined

It should be noted that it is not only preparations of mixed polysaccharides which differ from cellulose in solubility, but also esters (acetates, nitrates) do not dissolve completely in solvents of the corresponding cellulose esters. Thus, the nitrate of mixed polysaccharide (III), which contains 42 mol.-% of altrose, dissolves in acetone to 68%, and the triacetate dissolves in methylene chloride to 65%.

Thus, the data obtained permit us to conclude that the change in the configuration of the substituents and the conformation of the repeating unit has a considerable bearing on the structure and chemical and physicochemical properties of a polysaccharide.

II. Synthesis of Mixed Polysaccharides Containing Repeating Units of Amino Sugars and their N-Substituted Derivatives

As pointed out above, a mixed polysaccharide containing amino groups at the secondary carbon atoms of the repeating unit can be synthesized by reacting 2,3-anhydro derivatives of cellulose with ammonia. A mixed polysaccharide of this type can, however, also be prepared by the reaction of 2(3)-O-tosylcellulose with ammonia. This reaction proceeds with the intermediate formation and subsequent opening of α-oxide rings according to the following scheme:

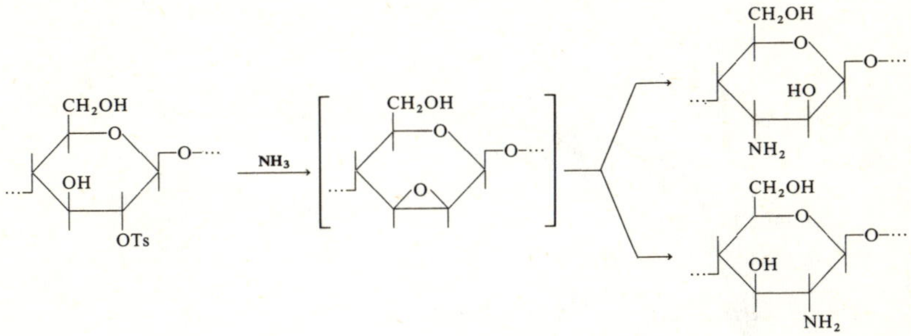

An investigation by paper chromatography of the products of hydrolysis of a mixed polysaccharide (28) indicated that the mixed polysaccharide contained, in addition to glucose units, repeating units of four amino sugars, two of which were identified as 3-deoxy-3-aminoaltrose and 2-deoxy-2-aminoglucose. The 3-aminoaltrose formed was found to be 80% of the total amino sugar content of the hydrolysate.

The replacement of the tosyloxy groups in cellulose tosylate and its derivatives upon interaction with amines leads to the formation of mixed polysaccharides containing units of N-alkyl(aryl)-substituted aminodeoxy sugars.

A study of the effect of the structure and size of the amine radical (butylamine, dibutylamine, hexylamine, piperidine, aniline) on its reactivity in the nucleophilic replacement of tosyloxy groups in cellulose tosylate has shown (29) that this reactivity depends primarily on the basicity of the amine. This assumption was confirmed by a study of the interaction of cellulose toxylates of different degrees of substitution (DS from 0.3–1.7) wit iminodiacetic and anthranilic acids (30, 31). The degree of substitution of the reaction products with respect to the N-carboxyalkyl(aryl)amino groups was found to be lower than that of the corresponding N-alkyl(aryl)amino derivatives.

The low nucleophilicity of the amino (imino) groups in iminodiacetic and anthranilic acids meant that the DS of the products of their interaction with cellulose tosylate did not exceed 0.7–0.8. Hence, the appearance of a product with a DS of about 2 through the interaction of anthranilic acid with cellulose nitrate was unexpected (32). The reaction rates for cellulose toxylate and nitrate are shown in Fig. 3.

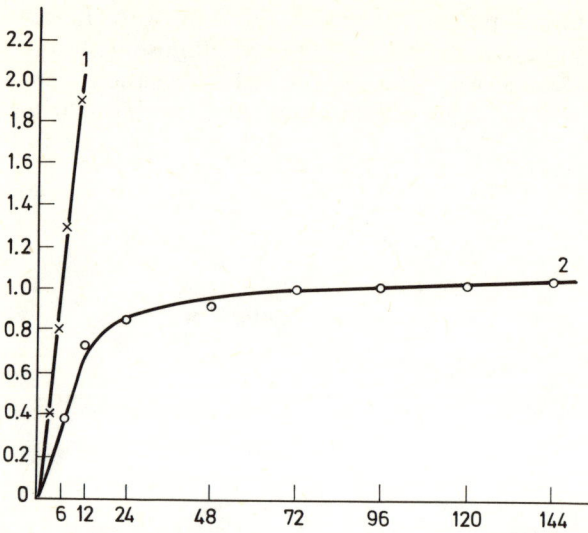

Fig. 3. Time dependence of the degree of substitution of products of the interaction of cellulose tosylate and nitrate with anthranilic acid in aqueous solution. Reaction time (in hours) is plotted along the abscissa, degree of substitution with respect to amino groups along the ordinate: *1* cellulose nitrate; *2* cellulose tosylate

III. Synthesis of Mixed Polysaccharides: C-Alkyl Derivatives of Deoxycellulose

The synthesis of C-alkyl derivatives of monosaccharides is based on methods involving the interaction of halodeoxy derivatives of monosaccharides with Grignard reagent. Reactions of this type, however, permit extension of the carbon skeleton only at the first carbon atom (*33*). As we have already shown (*34*), the replacement of a Grignard reagent by more reactive organometallic compounds, in particular, organolithium compounds, enables the synthesis of a new class of cellulosebased mixed polysaccharides containg units of 6-C-alkyl-6-deoxyglucose.

The reaction of butyllithium with cellulose iodotosylate prepared by iodination of cellulose tosylate of high DS (1.65) has been studied. General considerations concerning the direction of the reactions proceeding in the system $RX + R'\text{Li}$; suggest that the interaction between iododeoxycellulose and organolithium compounds may involve the "normal" Wurtz reaction (7), the reaction of metallization (8), and the formation of products of spatial structure as a result of the Wurtz intermolecular reaction between metallized units and iododeoxyglucose units (9):

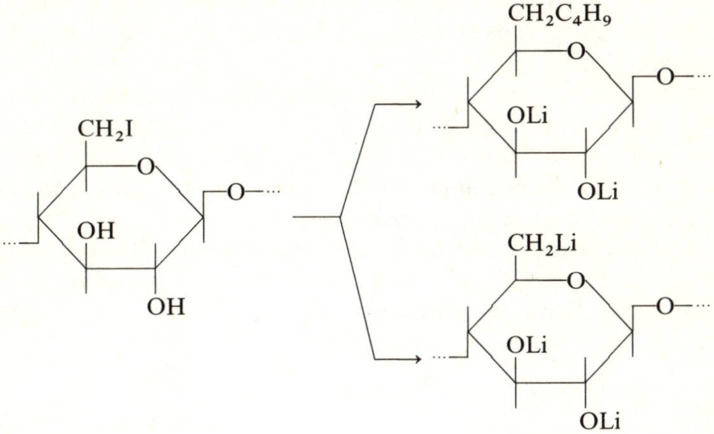

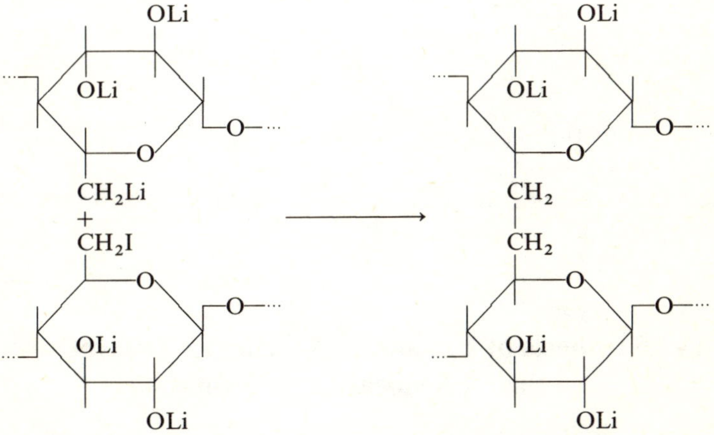

The reaction was carried out in a medium of petroleum ether, tetrahydrofuran (THF), and diethyl ether. According to the data obtained, the 6-C-butyl derivative of deoxycellulose forms only in a medium of solvating solvents (ether, THF). The reaction product had a DS of 1.1 with respect to butyl groups when the reaction was carried out in THF.

Because of the high reactivity of alkyllithium, the C-alkylation of cellulose is accompanied by the cleavage of acetal linkages. A change to less active organolithium compounds may lead not only to the synthesis of new classes of derivatives but also to a decrease in the intensity of destructive reactions. Of considerable interest here are the lithium derivatives of barenes, a new class of boron-containg organoelement compounds possessing a unusual set of properties: exclusively high stability of the barenic nucleus, high degree of delocalization of electrons, and ability to enter into substitution reactions proceeding by different mechanisms (35, 36).

The boron-containing cellulose esters prepared to date have been readily hydrolyzable, even in air (37). The use of lithium derivatives of barenes makes it possible to produce boron-containing derivatives of cellulose in which the boron atoms are stably bount to the carbon atoms of the repeating unit of the macromolecule.

The interaction of cellulose iodotosylate (DS = 0.85–1.0 with respect to iodine and DS = 0.34–0.53 with respect to tosyloxy groups) and also of monotosylate with phenylbarenyl lithium, proceeding by the sheme:

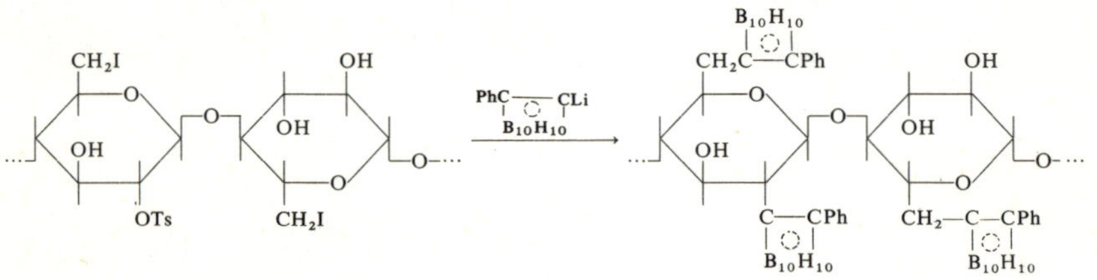

enabled the synthesis of phenylbarenyllithium derivatives of deoxycellulose with a DS of up to 1.32 (38).

IV. Synthesis of Unsaturated Cellulose Derivatives and their Chemical Transformations

The synthesis of unsaturated derivatives of cellulose is of interest because of the possibility of using the multiple bonds for further chemical transformations.

A number of methods can be employed: (1) synthesis of cellulose derivatives containing double bonds between the carbon atoms of the repeating unit can be effected by thermal decomposition of cellulose alkylxanthates and dehydroiodination of iododeoxy derivatives of cellulose; (2) synthesis of cellulose ethers and esters containing multiple bonds between the C atoms of the alkyl (acyl) radical.

Unsaturated derivatives of cellulose containing double bonds directly in the pyranose ring have been synthesized by the thermal decomposition of cellulose methylxanthate or bis-xanthate (39, 40), which proceeds as follows:

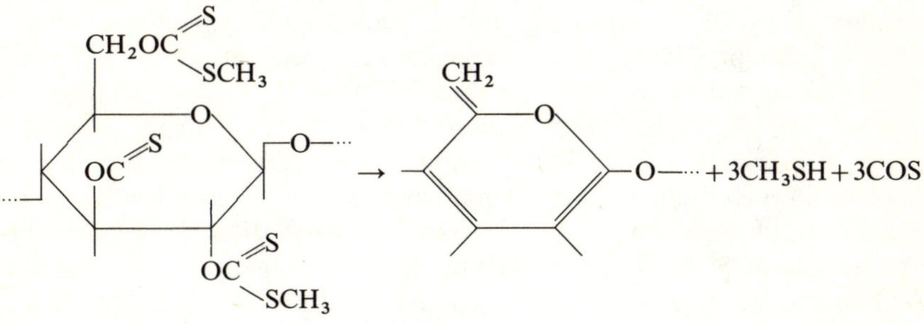

$$\rightarrow \quad +3CH_3SH + 3COS$$

The heating of cellulose trimethylxanthate at 160–180° C for 6 h resulted in the formation of a product containing 170–190 double bonds per 100 repeating units of the macromolecule.

For synthesis of another unsaturated derivative, the so-called 5,6-cellulosene, the method proposed was dehydroiodination of 6-iodo-*t*-deoxycellulose with absolute piperidine (41):

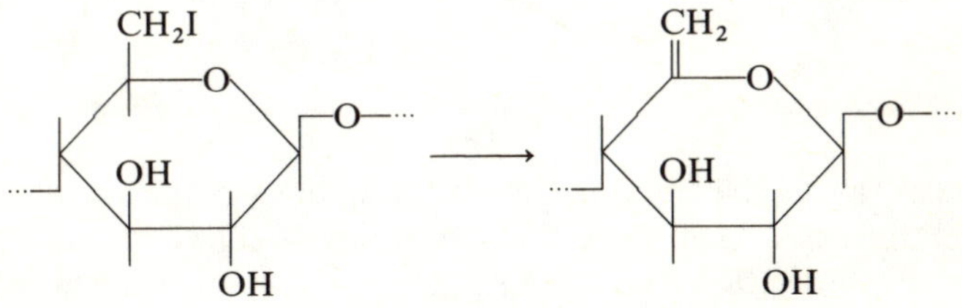

The principal reaction in this case is accompanied by the side reaction of nucleophilic substitution of iodine for piperidine groups.

A better method is to carry out dehydroiodination with an alcoholic solution of KOH, which yields a mixed polysaccharide containing 5,6-glucosene and 3,6-anhydroglucose units (42, 43).

A comparison of data on the monosaccharide composition of mixed polysaccharides formed by the action of a KOH solution on cellulose tosylate (see above) or 6-iododeoxycellulose indicates that the direction of the reactions taking place in the repeating units of the macromolecule under the action of nucleophilic reagents is strongly dependent on the nature of the leaving group. If the treatment of cellulose tosylate with an alcoholic solution of KOH results in almost quantitative formation of 3,6-anhydro rings, then two competing processes must occur in the treatment of 6-iododeoxycellulose: (1) formation of 3,6-anhydro rings (intramolecular nucleophilic substitution), and (2) formation of double bonds between $C_{(5)}$ and $C_{(6)}$ (elimination). The presence at the primary carbon atorm of the repeating unit of such a bulky substituent as the iodine atom probably hinders the change in the repeating unit conformation, which is necessary for the formation of 3,6-anhydro rings; as a result the probability of the elimination reaction is sharply increased.

A cellulose derivative with double bonds between the secondary carbon atoms of the repeating unit has been synthesized by dehydroiodination of preparations of 2(3)-iodo-2(3)-deoxycellulose prepared by the action of a solution of NaI in dimethylformamide on cellulose 6-O-trityl-2(3)-O-*p*-nitrobenzenesulphonate (44):

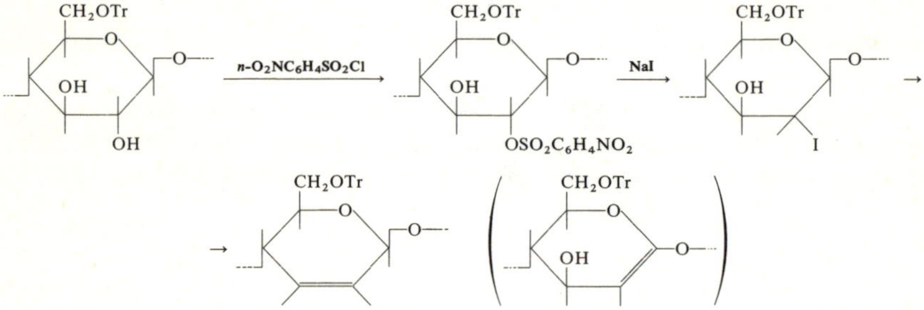

Because the double bond in the repeating units of the macromolecule of 5,6-cellulosene is polarized, this derivative of cellulose readily enters into chemical reactions proceeding by various mechanisms. In particular, the addition of methanol or acetic acid, when the reaction proceeds under mild conditions by an ionic mechanism, yields a mixed polysaccharide containing 5-O-methyl- or 5-O-acetylisoramnonose, respectively (43):

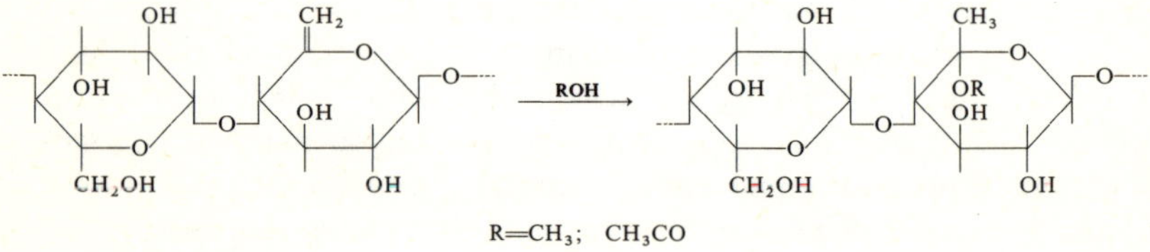

The interaction of 5,6-cellulosene with tributyl lead hydride during the formation of the latter gave a lead-containing derivative of 6-deoxy-cellulose (45):

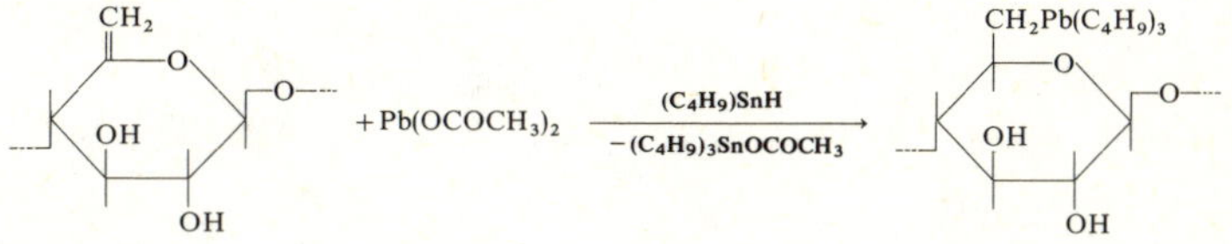

$$\text{[structure: 5,6-cellulosene]} + Pb(OCOCH_3)_2 \xrightarrow[-(C_4H_9)_3SnOCOCH_3]{(C_4H_9)SnH} \text{[structure: 6-deoxycellulose with CH}_2Pb(C_4H_9)_3\text{]}$$

The degree of conversion of doule bonds in the macromolecule of 5,6-cellulosene was about 50%.

Interesting possibilities for the synthesis of new types of polysaccharide derivatives are offered by the reaction of addition to double bonds, which proceeds by a free-radical mechanism. The presence of initiators of free-radical polymerization (benzoyl peroxide, tert-butyl peroxide, dicumyl peroxide, dinitrile of azodiisobutyric acid), also irradiation with ultraviolet light, has effected the addition to 5,6-cellulosene of chloroform, carbon tetrachloride, methylmonochloroacetate, dimethylphosphite and other compounds that decompose under the conditions of a reaction with the formation of free radicals (43, 46). The reaction proceeds as follows:

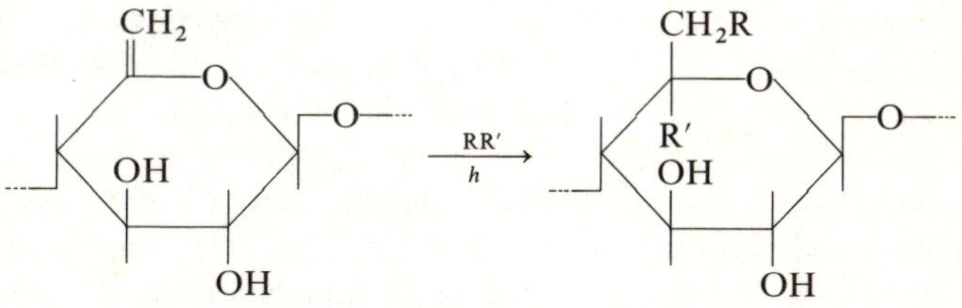

$$\text{[structure: 5,6-cellulosene]} \xrightarrow[h]{RR'} \text{[structure with CH}_2R\text{, R', OH]}$$

$R = -CCl_3; -CH_2COOCH_3; -\underset{\underset{O}{\|}}{P}(OCH_3)_2$ etc.

$R' = H; Cl$

The degree of conversion of double bonds was 65–75%. The main reaction route is the addition of a radical of higher molecular weight to the primary carbon atom of the repeating unit.

An analogous scheme was used (47) to synthesize a tin-containing derivative of 6-deoxycellulose: a preparation of 5,6-cellulosene swelled in benzene was treated with a solution of tri-n-butyltin hydride in absolute benzene at 80° C in the presence of azodiisobutyric dinitrile:

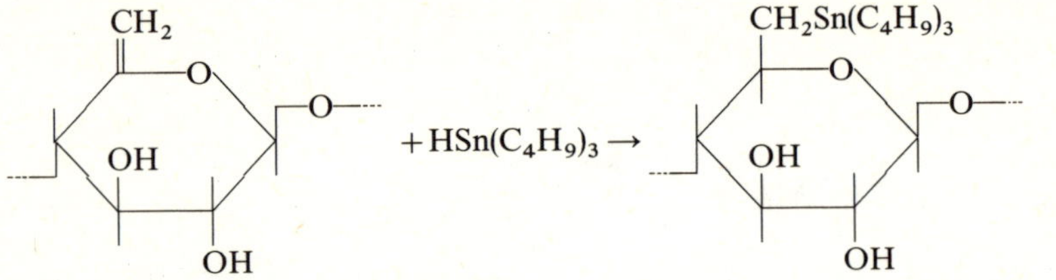

Apart from unsaturated cellulose derivatives with multiple bonds between the carbon atoms of the repeating unit, unsaturated cellulose ethers and esters are also of considerable interest.

There are relatively many reports in the literature of the synthesis of unsaturated cellulose ethers and esters containing double carbon–carbon bonds [e.g. Refs. (48) to (55)]. However, there are practically no data on the synthesis of cellulose derivatives containing triple carbon–carbon bonds: a single publication by Scherer and coworkers (56) describes the preparation of ethynyldeoxycellulose by the interaction of cellulose nitrate with sodium acetylide in liquid ammonia.

We have systematically studied the possibility of preparing various types of cellulose derivatives containing triple bond. Certain chemical transformations of the resulting compounds have also been accomplished.

The possibility of preparing a cellulose ester and the simplest acetylenecarboxylic acid-propiolic acid has been investigated. The data obtained (57) show that cellulose propiolates of high degree of substitution cannot be prepared by the reaction of cellulose with propiolyl chloride. The highest DS obtained was 0.12, probably because of the extreme instability of the acylating reagent used to effect hydrolysis.

Propiolic acid is a strong acid ($K = 1.4 \cdot 10^{-2}$), therefore one would expect the interaction between this acid and cellulose to involve the acylation of cellulose according to the scheme:

$$[C_6H_7O_2(OH)_3]_n + nx\,HC\equiv C.COOH \rightarrow$$
$$\rightarrow [C_6H_7O_2(OH)_{3-x}(OCOC\equiv CH)_x]_n + nx\,H_2O.$$

The esterification of cellulose by propiolic acid proceeds only in the presence of a catalyst and at a sufficiently high concentration of acid ($>80\%$).

The synthesis of cellulose propiolate has also been accomplished by way of a nucleophilic substitution reaction involving the interaction between cellulose tosylate and sodium propiolate in dimethylformamide according to the following scheme:

$$[C_6H_7O_2(OH)_{3-x}(OTs)_x]_n + nx\,HC{\equiv}C{-}COONa \rightarrow$$
$$\rightarrow [C_6H_7O_2(OH)_{3-x}(OCOC{\equiv}CH)_x]_n + nx\,TsONa.$$

The highest degree of substitution of the cellulose propiolate synthesized by this scheme was 0.58 (120° C, 40 h). The relatively low DS of the resulting esters is probably accounted for by the low nucleophilicity of the propioloxy ion, which is associated with the electron-accepting nature of a triple bond.

We have studied the possibility of using certain reactions, known for low-molecular-weight acetylene derivatives, to effect chemical transformations of cellulose propiolate. We have synthesized metal-containing derivatives, i.e. silver and copper acetylides of cellulose propiolate, having a silver content of up to 25% and a copper content of up to 15%. The reaction proceeds as follows:

$$[C_6H_7O_2(OH)_{3-x}(OCOC{\equiv}CH)_x]_n \xrightarrow{Ag^+} [C_6H_7O_2(OH)_{3-x}(OCOC{\equiv}CAg)_x]_n.$$

When cellulose propiolate (DS = 0.4 – 0.9) is acted on by an aqueous solution of mercuric acetate, hydration occurs and this leads to the quantitative formation of a cellulose ester and pyroracemic acid:

$$[C_6H_7O_2(OH)_{2.1}(OCOC{\equiv}CH)_{0.9}] \xrightarrow[Hg(OAc)_2]{H_2O} \left[C_6H_7O_2(OH)_{2.1}\left(\underset{\underset{O}{\|}}{OC}{-}\underset{\underset{O}{\|}}{C}{-}CH_3\right)_{0.9}\right]_n.$$

The nucleophilic addition of methanol across the triple bond in the repeating unit of the macromolecule of cellulose propiolate resulted in the synthesis of cellulose α-methoxyacrylate. This reaction also proceeds quantitatively.

The synthesis of a cellulose ether containing a triple C—C bond — propargylcellulose — was accomplished by O-alkylation through the interaction of alkali cellulose and cellulose sodium alcoholate with propargyl bromide (58). The reaction proceeds according to the following scheme:

$$[C_6H_7O_2(OH)_{3-x}(ONa)_x]_n + nx\,CH{\equiv}C{-}CH_2Br \rightarrow$$
$$\rightarrow [C_6H_7O_2(OH)_{3-x}(OCH_2C{\equiv}CH)_x]_n + nx\,NaBr.$$

The alkali cellulose was allowed to react with propargyl bromide at room temperature for 72 h and yielded propargylcellulose with a DS of 2.6. An ester with the maximal degree of substitution (DS = 3.0) was made by the action of propargyl bromide on cellulose trisodium alcoholate in liquid ammonia. The fact that is is possible to prepare propargylcellulose with a high degree of substitution under relatively mild conditions is explained by the high reactivity of the O-alkylating reagent (propargyl bromide) in which the halogen atom is activated due to the presence of an electron-accepting acetylene group.

Highly substituted preparations of propargylcellulose (DS > 2.5) that have not been dried will dissolve in dimethylformamide. After drying these preparations lose their solubility, probably due to oxidation in the air or crosslinking across the triple bonds.

Propargylcellulose containing reactive ethynyl groupings was used to prepare new types of cellulose esters: acetonylcellulose (1), 2-methoxypropen-1-ylcellulose (II), 2-diethylaminopropen-1-ylcellulose (III):

$$[C_6H_7O_2(OH)_{3-x}(OCH_2C{\equiv}CH)_x]_n \begin{array}{c} \xrightarrow{\underset{(Hg^{2+})}{H_2O}} \\ \xrightarrow{CH_3OH} \\ \xrightarrow{(C_2H_5)_2NH} \end{array} \begin{array}{l} \left[C_6H_7O_2(OH)_{3-x}\left(\begin{array}{c}OCH_2CCH_3\\ \|\\ O\end{array}\right)_x\right]_n \quad (I)\\ \left[C_6H_7O_2(OH)_{3-x}\left(\begin{array}{c}OCH_2C{=}CH_2\\ |\\ OCH_3\end{array}\right)_x\right]_n \quad (II)\\ \left[C_6H_7O_2(OH)_{3-x}\left(\begin{array}{c}OCH_2C{=}CH_2\\ |\\ N(C_2H_5)_2\end{array}\right)_x\right]_n \quad (III) \end{array}$$

The reaction of propargylcellulose with the reagents indicated proceeds quantitatively.

As reported in the literature (59), a propargylic ether of cellulose was used for the synthesis of tin-containing derivatives of cellulose. The action of bis(triethyltin) and bis(tri-n-butyltin) oxides on the propargylic ether of cellulose in absolute benzene at 80° C resulted in the synthesis of 3-(trialkylstannyl)propargylic ethers of cellulose.

The interaction of bis(trialkyltin) oxide with a propargylic ether of cellulose takes place only with previously activated preparations of the propargylic ether. The degree of conversion of the alkyl residues of the propargylic ether of cellulose is substantially affected by the basicity of the organotin compound used in the reaction. 3-(Triethylstannyl)-propargylic ether of cellulose is readily hydrolyzed when heated in water (1 h at 60° C), with the complete splitting-off of the organotin groups.

Organotin derivatives of a cellulose ether were also synthesized through the interaction between propargylcellulose and tributyltin hydride:

$$\text{Cell—OCH}_2\text{C}{\equiv}\text{CH} + \text{HSn}(\text{C}_4\text{H}_9)_3 \rightarrow \text{Cell—OCH}_2\text{CH}{=}\text{CH—Sn}(\text{C}_4\text{H}_9)_3.$$

The reaction proceeds only in the presence of azodiisobutyrodinitrile or on γ irradiation of the reaction mixture. The maximum degree of conversion of triple bonds was 72–77%.

Ethynyl- and phenylethynyldeoxycellulose have also been synthesized (60) through the interaction of cellulose tosylate with sodium acetylide and phenylacetylide in liquid ammonia according to the scheme:

$$[\text{C}_6\text{H}_7\text{O}_2(\text{OH})_{3-x}(\text{OTs})_x]_n + nx\text{NaC}{\equiv}\text{CR} \rightarrow$$
$$\rightarrow [\text{C}_6\text{H}_2\text{O}_2(\text{OH})_{3-x}(\text{C}{\equiv}\text{CR})_x]_n + nx\text{NaOTs},$$
$$\text{R}{=}\text{H}; \ \text{C}_6\text{H}_5.$$

Secondary as well as primary tosyloxy groups can be replaced under the conditions adopted for the reaction of nucleophilic substitution, which indicates the high nucleophilicity of sodium acetylide and phenylacetylide. The replacement of the alkyl hydrogen by an aryl radical increases the nucleophilicity of the corresponding metal acetylides; this finding is consistent with literature data (61).

The structure of certain preparations of ethynyldeoxycellulose (IV) has also been verified by a study of the composition of the products of condensation of IV with acetone. The reaction is as follows:

$$[\text{C}_6\text{H}_7\text{O}_2(\text{OH})_{3-x}(\text{C}{\equiv}\text{CNa})_x]_n + nx(\text{CH}_3)_2\text{CO} \rightarrow$$
$$\rightarrow \left[\text{C}_6\text{H}_7\text{O}_2(\text{OH})_{3-x}\left(\underset{\underset{\text{OH}}{|}}{\text{C}{\equiv}\text{C—C}(\text{CH}_3)_2}\right)_x\right]_n.$$

This reaction leads to new deoxycellulose derivatives that contain a residue of acetylenic alcohol. Like the reactions on addition of water, methanol, or diethylamine, the condensation reaction with acetone proceeds quantitatively.

V. Synthesis of Organoelement Derivatives of Cellulose

1. Synthesis of Silicon-Containing Cellulose Esters

This class of cellulose derivatives has attracted the attention of investigators because of their specific properties, primarily their increased hydrophobicity. There is, however, a serious drawback, typical of most silicon-containing cellulose esters prepared by the action of trialkylchlorosilanes, namely, the low stability of the C–O–Si bond to hydrolysis. One way of overcoming this difficulty is to increase the length of the organosilicon radical.

Silicon-containing cellulose esters of high hydrolytic stability have been synthesized by the reaction of cellulose with α-chloro-ω-trimethylsilylpolydimethylsiloxanes in pyridine (62):

$$\text{Cell—OH} + \text{Cl}\left[\begin{array}{c}\text{CH}_3\\|\\-\text{Si}-\text{O}\\|\\\text{CH}_3\end{array}\right]_n -\text{Si(CH}_3)_3 \xrightarrow{-\text{HCl}} \text{Cell—O}\left[\begin{array}{c}\text{CH}_3\\|\\\text{Si}-\text{O}\\|\\\text{CH}_3\end{array}\right]_n \text{Si(CH}_3)_3$$

$$n = 4 - 16.$$

The degree of substitution of the synthesized cellulose esters ranged from 0.2 (at $n = 16$) to 0.52 ($n = 4$). The hydrolytic stability of the C–O–Si bond increases with increasing n due to the shielding effect of the radicals. Thus, whereas more than 90% of the Si–O–C bonds are hydrolyzed in a preparation of trimethylsilylcellulose ($n = 0$) on boiling in water for 10 hours, the figure falls to 24% at $n = 16$ (Fig. 4). The hydrolytic stability of silicon-containing esters also increases when there are no Si–O–C bonds in the ester. A cellulose ester of this type has been prepared by the reaction of alkali cellulose with halomethyltrialkylsilanes (63):

$$\text{Cell—OH} + \text{ClCH}_2\text{SiR}_3 \xrightarrow{\text{NaOH}} \text{Cell—OCH}_2\text{SiR}_3 + \text{NaCl} + \text{H}_2\text{O}$$
$$R = \text{CH}_3; \text{C}_2\text{H}_5.$$

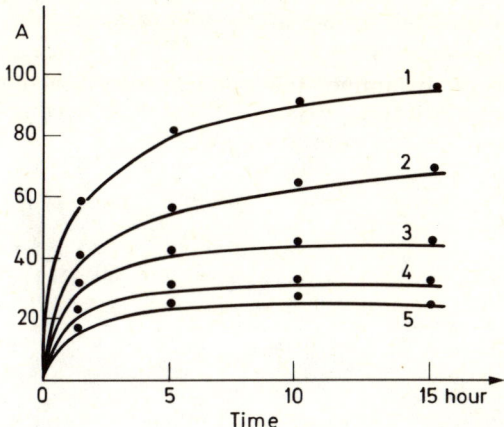

Fig. 4. Effect of the size of the organosilicon radical on the change of the stability of the Si—O—C bond in organosilicon derivatives of cellulose to the action boiling water. The number of dimethylsiloxy groups in the ester radical: *1* 1; *2* 4; *3* 8; *4* 12; *5* 16. The ordinate is the moisture absorption, per cent; the abscissa is the time, hours

Apart from the reaction with halogen derivatives, use was also made, for the synthesis of silicon-containing cellulose esters, of the reaction of alcoholysis, with cellulose, of tetraalkoxysilanes or alkyl(aryl)trialkoxysilanes and the amides of siliconic acids (*64*):

$$\text{Cell—OH} + CH_3Si(OC_4H_9)_3 \rightarrow \text{Cell—O—Si(OC}_4H_9)_2 + C_4H_9OH$$
$$\underset{CH_3}{|}$$

$$\text{Cell—OH} + (CH_3)_2Si[N(C_2H_5)_2]_2 \rightarrow \text{Cell—OSi(CH}_3)_2 + HN(C_2H_5)_2 \, .$$
$$\underset{N(C_2H_5)_2}{|}$$

2. Synthesis of Fluorine-Containing Cellulose Ethers and Esters

The preparation of fluorine-containing cellulose derivatives is of considerable scientific and practical interest because, due to the enhanced eletrophilicity of fluorine-containing substituents, the reactions occur under comparatively mild conditions. Moreover, the introduction of fluorine-containing groupings leads to materials of high hydro- and oleophobicity.

The synthesis of a low-substituted cellulose ester and α-hydroxyfluoroisobutyric acid (DS = 0.06–0.07) has been accomplished by acylation of cellulose with hexafluorodimethylketene

$$\text{Cell—OH} + \begin{array}{c} F_3C \\ \diagdown \\ F_3C \end{array} C=C=O \rightarrow \text{Cell—OCOCH} \begin{array}{c} CF_3 \\ \diagup \\ \diagdown \\ CF_3 \end{array}$$

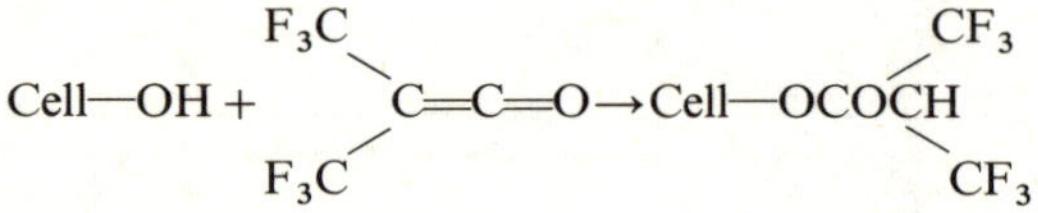

The low degree of substitution of a fluorine-containing cellulose ester stems from the difficulty with which the acylating reagent diffuses into the cellulosic fiber. Increased degree of substitution was achieved by carrying out the reaction in a homogeneous medium through the acylation, with hexafluorodimethylketene, of hydroxyl groups in incompletely substituted cellulose esters (secondary acetate, nitrate) and ethers (cyanoethyl). The structure of the products obtained was confirmed by infrared and ^{19}F nmr spectroscopy.

In the synthesis of fluorine-containing cellulose ethers we made use of nucleophilic addition of reagents of low nucleophilicity, such as alcohols, to the double bonds of fluorolefins, made possible by the high electronegativity of fluorine. The reaction yields mainly saturated α, α-difluoroethers, but is accompanied by a side reaction with formation of unsaturated ethers, that occurs as a result of the replacement of one of the fluorine atoms:

$$[C_6H_7O_2(OH_3)]_n + \overset{\delta+}{CF_2}=CF-CF_3 \rightarrow$$
$$\rightarrow [C_6H_7O_2(OH)_{3-x-y-z}(OCF_2CHFCF_3)_x(OCF=CF-CF_3)_y(OCF_2CF=CF_2)_y]_n.$$

The reaction of cellulose with tetrafluoroethylene, perfluoropropylene (66), and perfluoroisobutylene (67) has also been studied.

Cellulose does not react with tetrafluoroethylene and perfluoropropylene in the absence of a catalyst, but it does with perfluoroisobutylene, in which the specific properties of fluorolefins are especially pronounced due to the electron-accepting effect of the two unsymmetrically arranged trifluoromethyl groups, even in the absence of a catalyst. The fluorolefins under study can be arranged in the following order according to the reactivity exhibited in reactions with cellulose:

$$CF_2{=}CF_2 < CF_2{=}CF{-}CF_3 < CF_2{=}C\begin{smallmatrix}CF_3\\CF_3\end{smallmatrix}.$$

In the absence of a catalyst, cellulose also reacts with derivatives of perfluoroacrylic acid, i.e. perfluoroacrylonitrile and ethylperfluoroacrylate; these reactions give cellulose esters of a low degree of substitution (DS = 0.05–0.15).

3. Synthesis of New Types of Phosphorus-Containing Cellulose Esters

Studies of the synthesis of phosphorus-containing derivatives of cellulose fall into two groups: phosphorylation of cellulose with derivatives of phosphoric acids and phosphorous acids, respectively. Until recently, most studies have been of derivatives of cellulose and phosphoric compounds. There are a few publications reporting the synthesis of esters of phosphorous acids and cellulose, but the properties of these compounds have not been studied at all. However, esters of phosphorous acids and cellulose can be used for the preparation of esters of phosphoric acids and cellulose, which cannot be obtained by direct synthesis (or only with difficulty) by the reaction of the hydroxyl groups of the cellulose macromolecule with phosphorus-containing reagents.

The various ways in which esters of cellulose and phosphorous acids can be synthesized are: esterification of cellulose with free acids; alcoholysis, with cellulose, of the esters and amides of phosphoric acids; and esterification with mixed anhydrides of phosphoric acids and carboxylic acids.

Synthesis of Cellulose Hypophosphites

No data are available in the literature on the synthesis of high-molecular-weight compounds containing residues of hypophosphorous acid. Cellulose hypophosphites have been prepared by the action of hypophosphorous acid on cellulose (68) according to the scheme:

$$\text{Cell}{-}\text{OH} + \text{H}_3\text{PO}_2 \xrightarrow{-\text{H}_2\text{O}} \text{Cell}{-}\text{O}{-}\overset{H}{\underset{H}{P}}{=}\text{O}$$

Cellulose hypophosphites with DS = 0.95 have been produced by heating cellulose with the adsorbed acid in a current of an inert gas.

If the reaction time of the amount of hypophosphorous acid are increased, the phosphorus content of the cellulose hypophosphites will be increased. When the reaction temperature is raised from 80–120° C, a regular increase in degree of esterification is observed, but any further rise of temperature, other conditions being the same, causes the phosphorus content in the reaction product to decrease. This fact, unexpected at first sight, is probably explained by the partial decomposition of hypophosphorous acid to phosphine and phosphorous acid at temperatures above 120° C.

The cellulose esters obtained are only slightly stable to the action of hydrolyzing agents. For example, boiling for 30 min in distilled water causes the phosphorus content of a cellulose hypophosphite to fall from 4.6–1.2%; on boiling for one hour, it drops to 0.51%.

The structure of the cellulose hypophosphites synthesized has been confirmed by spectroscopy (69). The spectra of cellulose hypophosphites show bands in the regions 2380–2440 cm^{-1} and 1200–1250 cm^{-1}, assignable to the valence vibrations of the P–H and P–O groups, respectively.

For the synthesis of cellulose phosphites we used various methods, in particular esterification of a cellulose with phosphorous acid and alcoholysis, with cellulose, its esters and anhydrides.

Alcoholysis of the amides of alkylenephosphorous acids with cellulose led to the synthesis of hitherto unknown phosphorus-containing derivatives of cellulose (70), cellulose alkylenephosphites:

$$\text{Cell—OH} + (CH_3)_2\text{N—P}\begin{array}{c}\text{OCH}_2\\|\\\text{OCH}_2\end{array} \xrightarrow[-(CH_3)_2NH]{} \text{Cell—O—P}\begin{array}{c}\text{OCH}_2\\|\\\text{OCH}_2\end{array}$$

$$\text{Cell—OH} + (CH_3)_2\text{N—P}\begin{array}{c}\text{OCH}_2\\\diagdown\\\text{CH}_2\\\diagup\\\text{OCH}_2\end{array} \xrightarrow[-(CH_3)_2NH]{} \text{Cell—O—P}\begin{array}{c}\text{OCH}_2\\\diagdown\\\text{CH}_2\\\diagup\\\text{OCH}_2\end{array}$$

The rate of phosphorylation with the amide of propyleneglycolphosphorous acid is somewhat higher than with the amide of ethyleneglycolphosphorous acid. The treatment of cellulose with the cyclic dimethylamide of propyleneglycolphosphorous acid produced intermediate cellulose phosphites with a phosphorus content of up to 16% (DS = 1.92).

A study of the possibility of synthesizing cellulose phosphites of a high degree of substitution by trans-esterification of readily available dimethylphosphite has been reported in the literature (71). The transesterification of dimethylphosphite with cellulose may proceed in two ways:

(a) trans-esterification with replacement of one methoxyl group

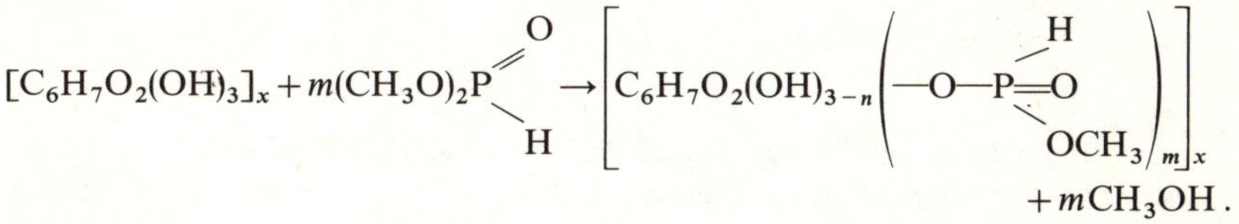

$$[C_6H_7O_2(OH)_3]_x + m(CH_3O)_2P\overset{O}{\underset{H}{\diagup}} \rightarrow \left[C_6H_7O_2(OH)_{3-n}\left(-O-P\overset{H}{\underset{OCH_3}{\diagdown}}=O\right)_m\right]_x + mCH_3OH.$$

(b) trans-esterification with replacement of two methoxyl groups

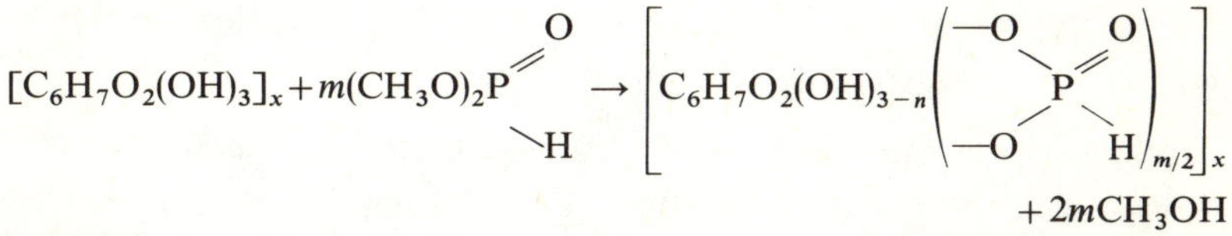

$$[C_6H_7O_2(OH)_3]_x + m(CH_3O)_2P\overset{O}{\underset{H}{\diagup}} \rightarrow \left[C_6H_7O_2(OH)_{3-n}\left(\overset{-O}{\underset{-O}{\diagdown}}P\overset{O}{\underset{H}{\diagup}}\right)_{m/2}\right]_x + 2mCH_3OH$$

The degree of substitution and the structure of the cellulose phosphites were determined by comparing the contents of phosphorus and methoxyl groups in the reaction products. The data (71) indicate that mainly mixed esters of methyl alcohol and cellulose with phosphorous acid form at the first reaction step; a longer reaction time results in almost complete replacement of the methoxyl groups according to scheme (b).

The phosphorylation of cellulose with dimethylphosphite was carried out with solvents of different polarity and basicity (dimethyl ether of diethyleneglycol, dimethylsulfoxide, dimethylformamide, dimethylaniline); the highest reaction rate between cellulose and dimethylphosphite, under identical reaction conditions occurs with dimethylformamide. In the presence of other solvents the rate of phosphorylation is much lower.

Cellulose phosphites have also been synthesized by the reaction of cellulose with monomethylphosphite (72), which could proceed as either a trans-esterification reaction

$$\text{Cell—OH} + \overset{CH_3O}{\underset{HO}{\diagdown}}P\overset{O}{\underset{H}{\diagup}} \xrightarrow{-CH_3OH} \text{Cell—O—}P\overset{OH}{\underset{H}{\diagup}}=O$$

or an esterification reaction

$$\text{Cell—OH} + \begin{array}{c}CH_3O\\ \diagdown\\ P\\ \diagup \diagup\\ HO \quad H\end{array} \!\!\!\!=\!\!O \xrightarrow{-H_2O} \text{Cell—O—P}\!\!=\!\!O \begin{array}{c}OCH_3\\ \\ H\end{array}$$

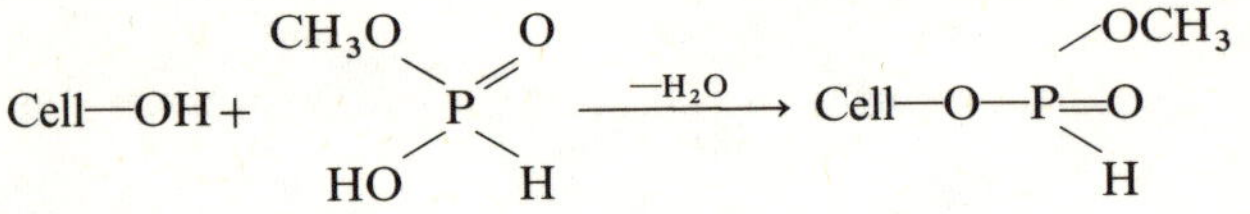

A comparison of the results of potentiometric titration of preparations of the starting cellulose and of the phosphorylated cellulose has shown that the latter contains acidic hydroxyl groups. It has also been found that all of the reaction products obtained contain methoxyl groups. These results, together with data from chromatographic studies, show that the phosphorylation of cellulose with monomethylphosphite involves the simultaneous occurrence of the esterification and transesterification reactions.

Despite wide variations of synthesis conditions, the treatment of cellulose with dimethylphosphite, monomethylphosphite and phosphorous acid failed to yield phosphites containing only one type of phosphorus-containing groups. This difficulty was overcome by using phosphoric anhydrides as the esterifying reagents. The action of mixed anhydrides of phosphorous acids and acetic acid on cellulose has yielded cellulose esters with alkyl-(methyl-)phosphorous acid (73):

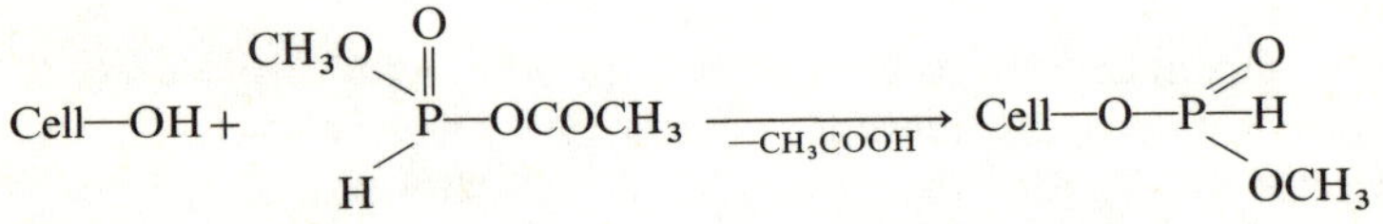

$$\text{Cell—OH} + \begin{array}{c}CH_3O\\ \diagdown\\ P\text{—OCOCH}_3\\ \diagup\\ H\end{array} \xrightarrow{-CH_3COOH} \text{Cell—O—P—H} \begin{array}{c}O\\ \\ OCH_3\end{array}$$

The phosphorylation of the previously activated cellulose was carried out at 50–60° C with an excess of a mixed phosphorus-containing anhydride, and also with a solution of the same anhydride in an organic solvent.

Normal cellulose phosphites were used for the synthesis of new types of phosphorus-containing derivatives of cellulose, i.e. cyclic phosphates (I), thionophosphates (II), thiolophosphates (III), and normal enol phosphates (IV):

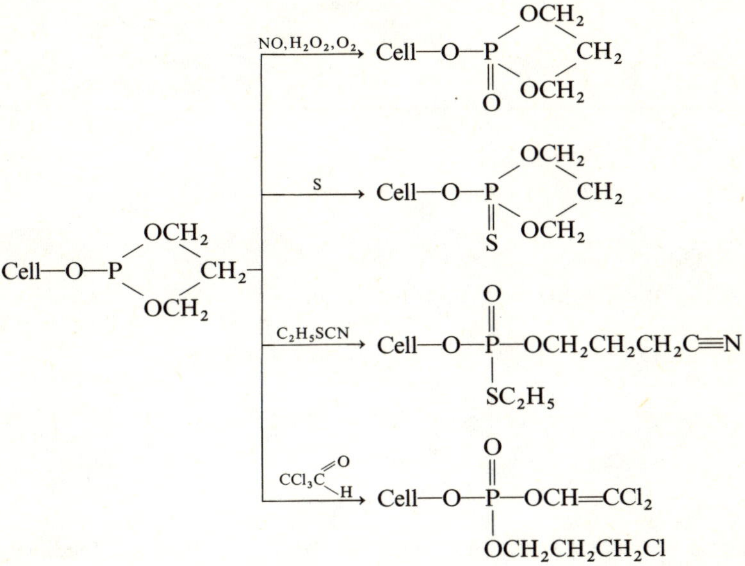

$$\text{Cell-O-P}\begin{array}{c}\text{OCH}_2\\ \diagup\quad\diagdown\\ \quad\quad\text{CH}_2\\ \diagdown\quad\diagup\\ \text{OCH}_2\end{array}$$

$\xrightarrow{\text{NO, H}_2\text{O}_2, \text{O}_2}$ Cell—O—P(=O)(OCH$_2$–CH$_2$–OCH$_2$) (I)

$\xrightarrow{\text{S}}$ Cell—O—P(=S)(OCH$_2$–CH$_2$–OCH$_2$) (II)

$\xrightarrow{\text{C}_2\text{H}_5\text{SCN}}$ Cell—O—P(=O)(SC$_2$H$_5$)—OCH$_2$CH$_2$CH$_2$C≡N (III)

$\xrightarrow{\text{CCl}_3\text{C(=O)H}}$ Cell—O—P(=O)(OCH=CCl$_2$)—OCH$_2$CH$_2$CH$_2$Cl (IV)

The conversion of acidic cellulose phosphites prepared from dimethyl- and monomethylphosphites permitted the synthesis of certain types of phosphorus-containing cellulose esters:

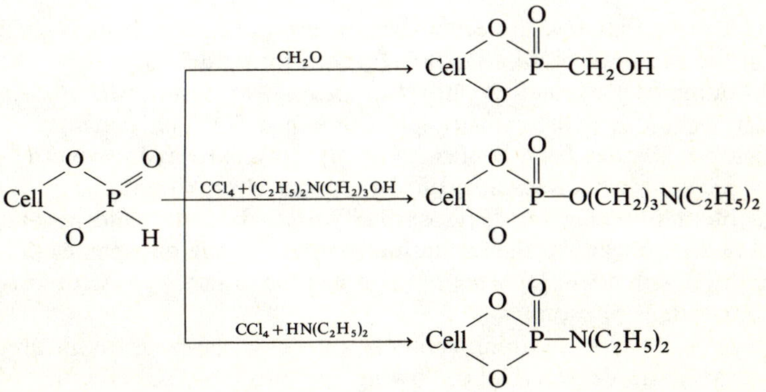

Cell—O—P(=O)(O)(H)

$\xrightarrow{\text{CH}_2\text{O}}$ Cell—O—P(=O)(O)—CH$_2$OH (V)

$\xrightarrow{\text{CCl}_4 + (\text{C}_2\text{H}_5)_2\text{N(CH}_2)_3\text{OH}}$ Cell—O—P(=O)(O)—O(CH$_2$)$_3$N(C$_2$H$_5$)$_2$ (VI)

$\xrightarrow{\text{CCl}_4 + \text{HN(C}_2\text{H}_5)_2}$ Cell—O—P(=O)(O)—N(C$_2$H$_5$)$_2$ (VII)

Table 10. Effect of the structure of phosphorus-containing acids on the fire-resistance of their cellulose esters

No.	Structure of Cellulose ester	Composition of Cellulose ester		Fire Resistance	
		P (%)	Value of γ	Angle of inflamability	Noncombustibility
1.	Cell$\diagup^O\diagdown_O\!\!P\diagup^{\!\!=\!O}_{\!\!H}$	2.87	34	180°	High
2.	Cell$\diagup^O\diagdown_O\!\!P(\!=\!\!O)\!-\!CH_2OH$	6.82	86	150°	Medium
3.	Cell$\diagup^O\diagdown_O\!\!P\diagup^{\!\!=\!O}_{\!\!OH}$	2.47	27.1	160°	Above medium
4.	Cell$\diagup^O\diagdown_O\!\!P(\!=\!\!O)\!-\!N(C_2H_5)_2$	5.81	80	150°	Medium
5.	Cell$\diagup^O\diagdown_O\!\!P(\!=\!\!O)\!-\!O(CH_2)_3N(C_2H_5)_2$	3.87	51.8	90°	Low

It is known that low-molecular-weight esters of phosphorous acid react, in the presence of alkali catalysts, with formaldehyde to give the corresponding hydroxymethyl phosphonates. This reaction was used to synthesize cellulose hydroxymethylphosphonates (V). The structure of these compounds has been confirmed by hydrolysis to hydroxymethylphosphonic acid, which was identified by paper chromatography.

The interaction of cellulose phosphites with carbon tetrachloride and alcohol or amine leads to the formation of intermediate phosphates (VI) and amidophosphates (VII) of cellulose in just the same way as with low-molecular-weight phosphites.

The phosphorus-containing cellulose esters synthesized are all fire-resistant to some degree. Data showing the effect of the structure of phosphorus-containing acids and their derivatives on the fire resistance[1] of the cellulose esters obtained are given in Table 10.

4. Synthesis of Tin-Containing Cellulose Esters

We have already described some ways of synthesizing tin-containing derivatives of cellulose through chemical transformations of 5,6-cellulosene and propargylic ethers of cellulose. These methods are based on the interaction of tin-containing compounds with multiple carbon–carbon bonds of cellulose derivatives. Tin-containing cellulose derivatives have also been synthesized by reacting bis-(tributyltin) oxide with cellulose esters containing carboxylic groups (carboxymethylcellulose) and mercapto groups (e.g. 2-hydroxy-3-thiolopropylcellulose) (59) according to the following scheme:

$$\text{Cell—OCH}_2\text{COOH} + [(C_4H_9)_3\text{Sn}]_2\text{O} \rightarrow \text{Cell—OCH}_2\text{C}\begin{smallmatrix}\diagup\!\!\!\!O\\ \diagdown\text{OSn}(C_4H_9)_3\end{smallmatrix}$$

The tin content of the esters synthesized was up to 19% (DS up to 0.6). In the reaction with bis-(tributyltin) oxide the carboxylic groups are more reactive than the thiol groups.

Tin-containing esters have also been synthesized by the reaction of the sodium salt of carboxymethylcellulose with tributyltin chloride. The reaction proceeds as follows:

$$\text{Cell—OCH}_2\text{COONa} + \text{ClSn}(C_4H_9)_3 \rightarrow \text{Cell—O—CH}_2\text{COOSn}(C_4H_9)_3.$$

The tin content of the tributylstannylcarboxymethyl cellulose ester is practically unchanged on boiling in water for one hour. When subjected to hydrolysis with 0.1 N solutions of HCl and NaOH for 1 h at 20° C, the tributylstannyl residues of this ester are completely split off. Thus, polymeric acylates of trialkyltin, like low-molecular-weight organotin compounds of this type, are unstable to the action of aqueous solutions of acids and alkalis.

---

[1] Fire resistance was determined by the procedure indicated in Ref. (74).

VI. Discussion of the Principles Governing the Synthesis of Cellulose Esters by the Trans-esterification Reaction

Among the various types of cellulose derivatives, it is the cellulose esters that find the widest practical application and complete succesfully in a number of branches of the industry with synthetic polymers. The complete or partial esterification of the hydroxyl groups of cellulose can yield cellulosic materials having such technically valuable properties as thermoplasticity, hydrophobicity, resistance to heat and light, stability to the action of putrefying microorganisms, bactericidal action, etc.

Of the methods of synthesis of cellulose esters, the one that has been most thoroughly studied is the reaction of trans-esterification, and this method is widely used for the synthesis of low-molecular-weight esters. The alcoholysis of a low-molecular-weight ester (methyl- and n-propylborate) with hydroxyl groups of cellulose was first used (37) for the preparation of cellulose borate. This was followed by the trans-esterification, with cellulose, of the esters of phosphorous acids (see above), i.e. mono-, di- and trimethylphosphites (71, 72, 75), esters of phosphonic acids (76), and also phenyl-β-chloroethyl- and β-fluoroethylphosphites (77, 78). Of considerable interest is the reaction of alcoholysis, with cellulose, of the esters of aryl- and naphthalenesulphonic acid, which results in the formation of cellulose ethers, rather than esters (79–81).

No systematic studies of the principles governing the synthesis of cellulose esters by the trans-esterification reaction had been available in the literature until the present authors published work permitting one to ascertain the relation between the reactivity of low-molecular-weight esters and their structure, and the direction of the reactions involved.

The effect of the degree of polarization of the ester linkages in low-molecular-weight esters on their reactivity has been examined by studying the reaction of alcoholysis, with cellulose, of methyl esters of benzoic, p-chloro-, p-hydroxy, p-hydroxy-, p- and o-nitrobenzoic acids, and those of phenyl-, phenoxy-, 2,4-dichlorophenoxy- and monochloroacetic acids. The trans-esterification reaction was carried out in nonaqueous dimethylformamide in the presence of catalysts (sodium methylate, cadmium acetate, p-toluenesulphonic acid) at 110–140° C.

The experimental evidence obtained has shown that the reactivity of the esters of the aromatic and aliphatic carboxylic acids under study in the reaction of alcoholysis with cellulose changes substantially with changes in the chemical constitution of the acyl radical. According to existing conceptions (82), the alcoholysis of esters in the presence of an alkaline catalyst at the rate-determining step involves the attack of the alkoxy ion on the carbon atom of the carbonyl group; in the case of acid catalysis, protonation of the carbonyl oxygen atom is followed by the attack of the OH group of the alcohol on the carbonyl carbon atom.

$$CH_3-O-\underset{\parallel}{\overset{O}{C}}-R \overset{H^+}{\rightleftarrows} CH_3-O-\underset{\underset{Cell-\ddot{O}H}{\uparrow}}{\overset{OH}{\overset{|}{C^+}}}-R \rightleftarrows CH_3O-\underset{\underset{Cell-O^+H}{|}}{\overset{OH}{\overset{|}{C}}}-R \rightleftarrows$$

$$\rightleftarrows CH_3\overset{+}{O}-\underset{\underset{OCell}{|}}{\overset{OH}{\overset{|}{\underset{H}{C}}}}-R \xrightarrow{-CH_3OH} R-\underset{\underset{OCell}{|}}{\overset{O}{\overset{\parallel}{C}}} +H^+$$

$$H_3\overset{\delta+}{C}-O-\underset{\underset{Cell-\ddot{O}H}{\uparrow}}{\overset{O^{\delta-}}{\overset{\parallel}{C}}}-R \rightleftarrows H_3\overset{\delta+}{C}-O \cdot \overset{OH}{\overset{|}{C^+}}-R \rightarrow H_3C \underset{\underset{\underset{Cell-OCH_3+H^+}{\downarrow}}{CellO^+:H}}{\overset{OH}{\overset{|}{\underset{\parallel}{|}}}} +\underset{\parallel}{\overset{|}{C}}-R$$

Therefore the reactivity of an ester must depend on the basicity of the carbon atom of the carbonyl group, which is associated with the electron-accepting effect of the acid radical.

The overall effect of the radicals of the substituents can be assessed by comparing the dissociation constants of the corresponding substituted acids data on the degree of substitution of various cellulose esters prepared by the trans-esterification reaction are presented in Table 11.

As can be seen from the data, the degree of polarization of an ester linkage, which is characterized by the dissociation constant of the acid whose acyl radical enters into the composition of a low-molecular-weight ester, has an effect both on the degree of substitution of the cellulose ester obtained and on the mechanism of alcoholysis. The alcoholysis of esters of weak carboxylic acids proceeds with the cleavage of only the acyl-oxygen bond and the formation of a cellulose ester. With increasing dissociation constant of the acid, the DS of the cellulose ester prepared by *trans*-esterification under identical conditions passes through the maximum situated in the region of $pK_a \approx 2-2.5$; in this case the mechanism of alcoholysis of the ester with cellulose is partially changed: the formation of an ester is accompanied by the O-alkylation of the hydroxyl groups of cellulose.

Table 11. Relation between the dissociation constants of acids contained in esters subjected to alcoholysis with cellulose and the DS of the resulting cellulose esters

Acid contained in ester		DS of cellulose ester	
R in RCOOH	Dissociation constant, $K_{25} \cdot 10^5$	With respect to acyl groups	With respect to alkyl groups
$C_6H_5CH_2$	4.88	0.45	—
C_6H_5	6.30	0.47	—
$p\text{-}ClC_6H_4$	10.3	0.54	—
$p\text{-}HOC_6H_4$	26.2	0.67	—
$p\text{-}O_2NC_6H_4$	37.2	0.59	—
$C_6H_5OCH_2$	75.6	0.64	—
$ClCH_2$	$1.38 \cdot 10^2$	0.74	0.06
$2,4\text{-}Cl_2C_6H_3OCH_2$	$2.30 \cdot 10^2$	0.87	0.15
$o\text{-}O_2NC_6H_4$	$6.75 \cdot 10^2$	0.54	0.18
$2,4,6\text{-}(O_2N)_3C_6H_2$	$22.4 \cdot 10^3$	0.48	0.75
$CF_3(CF_2)_2$	$68 \cdot 10^3$	0.01	1.70

A further increase in the degree of polarization of the bond results in an abrupt change of the mechanism of alcoholysis of an ester with cellulose. For example, methylperfluorobutyrate (K of the acid $= 6.8 \cdot 10^{-1}$) undergoes alcoholysis with the cleavage of practically only the alkyl–oxygen bond and formation of methyl ether of cellulose. From the data given, it is clear that the overall reactivity of the low-molecular-weight ester used as the acylating and O-alkylating reagent for cellulose increases with increasing dissociation constant of the carboxylic acid contained in the ester subjected to alcoholysis.

In certain cases, however, on alcoholysis, with cellulose, of methyl-p-hydroxy-, methyl-o-nitro- and methyl-2,4,6-trinitrobenzoates, the DS of cellulose esters differs from the value expected on the basis of the strength of the corresponding acid. Evidently, the degree of substitution of the cellulose esters obtained by the reaction of trans-esterification depends not only on the degree of polarization of the ester bond but on other factors as well. For example, the relatively high reactivity of methyl-p-hydroxybenzoate is accounted for by the increased swelling of cellulose in this reagent (the reaction proceeds at practically the same velocity in the absence of dimethylformamide, a solvent that provides in other cases the retention of the high degree of swelling of cellulose). The methyl esters of o-nitro- and 2,4,6-trinitrobenzoic acid are characterized by the decreased reactivity of cellulose in the reaction of acylation. This is porbably explained by the presence in the ortho-position to the carbonyl group of bulky substituents, which hinder sterically the interaction of the carbon atom of the carbonyl group with the OH groups of the cellulose macromolecule.

The principles established can probably also be applied to the reactions of alcoholysis, with cellulose, of the esters of acids containing other types of ionic groups. The statement that alcoholysis with cellulose proceeds with the cleavage of the alkyl–oxygen linkage and formation of cellulose ethers holds for the esters of sulfuric and arylsulfonic acids (*79–81*), i.e. for esters of strong acids. On the other hands, alcoholysis, with cellulose, of esters of weak acids, e.g. boric (*37*) and phosphorous acids (*70–78*), and also of phenyl methylxanthate and methylxanthoacetate (*83*), involves the cleavage of only the oxygen–radical linkage and the formation of cellulose esters.

VII. References

1. Rogovin, Z. A.: Chemical transformations and modification of cellulose. Moscow 1967 (in Russian).
2. — The chemistry of cellulose. Moscow 1972 (in Russian).
3. Lifshits, R. M., Rogovin, Z. A.: Usp. Khim. **6**, 1087 (1965); Progr. Polymer Chem. Nauka **1969**, 158 (in Russian).
4. Gal'braikh, L. S., Rogovin, Z. A.: In: Bikales, N. M., Segal, L. (Ed.): Cellulose and cellulose derivatives, part V, p. 877. Wiley-Interscience 1971.
5. Kochetkov, N. K., Bochkov, A. F., Dmitriev, B. A., Usov, A. I., Chizhov, O. S., Shibaev, V. N.: Khimiya Uglevodov (The chemistry of carbohydrates) Moscow 1967, p. 162 (in Russian).
6. Smirnova, G. N., Gal'braikh, L. S., Polyakov, A. I., Rogovin, Z. A.: Vysokomolekul. Soedin. **8**, 1396 (1966).
7. — — — Khim. Prirodn. Soedin. **2**, 3 (1966).
8. Robertson, J., Griffith, C. F.: J. Chem. Soc. **1935**, 1193
9. Belyakova, M. K., Gal'braikh, L. S., Rogovin, Z. A.: Vysokomolekul. Soedin. B **12**, 790 (1970).
10. Gal'braikh, L. S., Polukhina, S. I., Belyakova, M. K.: Makromol. Chem. **122**, 38 (1969).
11. Makhsudov, Yu. M., Gal'braikh, L. S., Polyakov, A. I., Rogovin, Z. A.: Vysokomolekul. Soedin. **8**, 1289 (1966).
12. — — Rogovin, Z. A.: Vysokomolekul. Soedin A **9**, 1733 (1967).
13. — Krylova, L. G., Gal'braikh, L. S., Rogovin, Z. A.: Khim. Prirodn. Soedin. **2**, 372 (1966).
14. Smirnova, G. N., Gal'braikh, L. S., Rogovin, Z. A.: Cellulose Chem. Technol. **1**, 11 (1967).
15. Makhsudov, Yu. M., Komar, V. P., Zhbankov, R. G., Krylova, L. G., Gal'braikh, L. S., Rogovin, Z. A.: Vysokomolekul. Soedin. **8**, 2012 (1966).
16. Barker, S. A., Stephens, R.: J. Chem. Soc. **1954**, 4550.
17. Foster, A. V., Stacey, M.: Acta Chem. Scand. **12**, 1819 (1958).
18. Gardner, S., Purves, C. B.: J. Am. Chem. Soc. **65**, 444 (1943).
19. Voitenko, I. L., Gal'braikh, L. S., Rogovin, Z. A.: Vysokomolekul. Soedin. B **13**, 66 (1971).
20. Tkacheva, L. P., Akim, E. L., Gal'braikh, L. S.: Vysokomolekul. Soedin. A **13**, 1819 (1971).
21. Reeves, R. E.: Advan. Carbohydr. Chem. **6**, 107 (1951).
22. Derevitskaya, V. A., Smirnova, G. S., Rogovin, Z. A.: Doklady Akad. Nauk SSSR **141**, 1090 (1961).
23. Gal'braikh, L. S., Voitenko, I. L., Rogovin, Z. A.: Cellulose Chem. Technol. **6**, 627 (1972).
24. Tkacheva, L. P., Gal'braikh, L. S., Frenkel, G. G., Rogovin, Z. A.: Vysokomolekul. Soedin. B **13**, 57 (1971).
25. Haworth, W. N., Owen, J. N., Smith, T.: J. Chem. Soc. **1941**, 88.
26. Voitenko, I. L., Kurdyukova, L. Ya., Bondarenko, O. A., Fainberg, E. Z., Gal'braikh, L. S., Rogovin, Z. A.: Vysokomolekul. Soedin. B **14**, 66 (1972).

27. Wolfrom, M. L., Taha, M. I., Horton, D.: J. Org. Chem. **28**, 3553 (1963).
28. Polukhina, S. I., Gal'braikh, L. S., Rogovin, Z. A.: Vysokomolekul. Soedin. B **11**, 270 (1969).
29. Kholmuradov, N., Kozlova, Yu. S., Gal'braikh, L. S., Rogovin, Z. A.: Vysokomolekul. Soedin. **8**, 1089 (1966).
30. Chaikina, E. A., Gal'braikh, L. S., Rogovin, Z. A.: Vysokomolekul. Soedin. **7**, 2014 (1965).
31. — Zholybina, L. F., Gal'braikh, L. S., Rogovin, Z. A.: Zh. VKhO im. Mendeleyeva **12**, 595 (1967).
32. — Gal'braikh, L. S., Rogovin, Z. A.: Vysokomolekul. Soedin. B **9**, 151 (1967).
33. Zhdanov, Yu. A., Dorofeenko, G. N.: The chemical changes of the carbon skeleton of carbohydrates. Ed. Akad. Nauk SSSR, Moscow 1962 (in Russian).
34. Akovbyan, E. M., Okhlobystin, O. Yu., Gal'braikh, L. S., Rogovin, Z. A.: Vysokomolekul. Soedin. **8**, 959 (1965).
35. Zakharkin, A. I., Stanko, V. I., Bratsev, V. A., Chakovsky, Yu. A., Okhlobystin, O. Yu.: Izv. Akad. Nauk SSSR, Ser. Khim. **1963**, 2233.
36. Heying, T. L., Ager, J. W., Klark, S. L. et al.: Inorg. Chem. **2**, 1089 (1963).
37. Makarov-Zemlyansky, Ya. Ya., Gertsev, V. V.: Zh. Obshch. Khim. **35**, 272 (1965).
38. Akovbyan, E. M., Chaikina, E. A., Bregadze, V. I., Okhlobystin, O. Yu., Gal'braikh, L. S., Rogovin, Z. A.: Vysokomolekul. Soedin. A **10**, 428 (1968).
39. Polyakov, A. I., Derevitskaya, V. A., Rogovin, Z. A.: Vysokomolekul. Soedin. **2**, 346 (1960).
40. — — — Vysokomolekul. Soedin. **5**, 161 (1963).
41. Lopatenok, A. A.: Dissertation, Leningrad, 1953.
42. Dimitrov, D. G., Antonyuk, L. S., Achval, W. B., Gal'braikh, L. S., Rogovin, Z. A.: Vysokomolekul. Soedin. A **10**, 1373 (1968).
43. — Gal'braikh, L. S., Rogovin, Z. A.: Cellulose Chem. Technol. **2**, 375 (1968).
44. Nikologorskaya, L. G., Kozlova, Yu. S., Gal'braikh, L. S., Rogovin, Z. A.: Vysokomolekul. Soedin. A **12**, 2762 (1970).
45. Nazar'ina, L. A., Okhlobystin, O. Yu., Rogovin, Z. A.: Vysokomolekul. Soedin. B **12**, 459 (1970).
46. Dimitrov, D. G., Gal'braikh, L. S., Rogovin, Z. A.: Vysokomolekul. Soedin. B **9**, 685 (1967).
47. Artemova, Yu. V., Virnik, A. D., Zemlyansky, N. N., Rogovin, Z. A.: Cellulose Chem. Technol. **5**, 319 (1971).
48. Sakurada, H.: Angew. Chem. **42**, 549 (1929).
49. Shostakovsky, M. F., Prilezhaeva, E. N., Tsymbal, L. V.: Zh. Obshch. Khim. **26**, 739 (1965).
50. Sharkova, E. F., Virnik, A. D., Rogovin, Z. A.: Izv. VUZ'ov SSSR, Khim. khim. tekhnol. **8**, 465 (1965).
51. Frank, G., Mendrzyk, H.: Ber. Deut. Chem. Ges. **63**, 875 (1930).
52. Berlin, A. A., Makarova, T. A.: Zh. Obshch. Khim. **21**, 1267 (1961).
53. Faraone, G.: J. Appl. Polymer Sci. **5**, 16 (1961).
54. Ioannidis, O. K., Aikhodzhaev, B. I., Pogosov, Yu. L.: Uzb. Khim. Zh., No. **4**, 76 (1964).
55. Engelman, H., Exner, F.: Makromol. Chem. **23**, 233 (1957).
56. Scherer, G. S., Saul, G. A.: Rayon Text. Monthly **28**, 474, 537 (1947).
57. Masaidova, G. S., Gal'braikh, L. S., Rogovin, Z. A., Nikitin, V. I.: Vysokomolekul. Soedin. A **9**, 1966 (1967).
58. — Yakunina, A. S., Gal'braikh, L. S., Rogovin, Z. A.: Vysokomolekul. Soedin. **8**, 865 (1966).

59. Artemova, Yu. V., Virnik, A. D., Zemlyansky, N. N., Rogovin, Z. A.: Cellulose Chem. Technol. **5**, 319 (1971).
60. Gal'braikh, L. S., Masaidova, G. S., Rogovin, Z. A.: Cellulose Chem. Technol. **3**, 455 (1969).
61. Nieuwland, J. A., Vogt, R. R.: The chemistry of acetylene. New York: Reinhold Publishing Corp. 1945.
62. Ivanov, N. V., Rogovin, Z. A., Andrianov, K. A.: Vysokomolekul. Soedin. Cellulose and its derivatives, 1963, p. 44.
63. — — Nguen Vin Chiy: Izv. VUZ'ov SSSR, Khim. Khim. tekhnol., No. **1**, 124 (1965).
64. Predvoditelev, D. A., Rogovin, Z. A.: Vysokomolekul. Soedin. **9**, 611 (1967).
65. Sletkina, L. S., Bargamova, M. D., Rogovin, Z. A.: Vysokomolekul. Soedin. Cellulose and its derivatives, Ed. Akad. Nauk SSSR 1963, p. 55.
66. — Rogovin, Z. A.: Cellulose Chem. Technol. **1**, 641 (1967).
67. — — Vysokomolekul. Soedin. B **9**, 348 (1967).
68. Predvoditelev, D. A., Nifantyev, E. E., Rogovin, Z. A.: Vysokomolekul. Soedin. **7**, 791 (1965).
69. Garbuz, N. I., Zhbankov, R. G., Predvoditelev, D. A., Nifantyev, E. E., Rogovin, Z. A.: Vysokomolekul. Soedin. **8**, 603 (1966).
70. Predvoditelev, D. A., Nifantyev, E. E., Rogovin, Z. A.: Vysokomolekul. Soedin. **8**, 213 (1966).
71. — Tyuganova, M. A., Nifantyev, E. E., Rogovin, Z. A.: Zh. VKhO im. Mendeleyeva **10**, 459 (1965).
72. — Nifantyev, E. E., Rogovin, Z. A.: Vysokomolekul. Soedin. **7**, 1005 (1965).
73. — — — Vysokomolekul. Soedin. **8**, 76 (1966).
74. Reeves, W. A., McMillan, O., Guthrie, J. D.: Textile Res. J. **23**, 527 (1953).
75. U Mey-Yan, M. A., Tyuganova, M. A., Gefter, E. L., Rogovin, Z. A.: Vysokomolekul. Soedin. Cellulose and its derivatives, Ed. Akad. Nauk SSSR 1963, p. 37.
76. Petrov, K. A., Nifantyev, E. E.: Vysokomolekul. Soedin. **4**, 242 (1962).
77. Predvoditelev, D. A., Nifantyev, E. E., Rogovin, Z. A.: Zh. prikl. Khim. **40**, 413 (1967).
78. Petrov, K. A., Nifantyev, E. E., Sopikova, I. I., Merkulova, M. I.: Vysokomolekul. Soedin. Cellulose and its derivatives, Ed. Akad. Nauk SSSR 1963, p. 86.
79. Plisko, E. A.: Zh. Obshch. Khim. **28**, 3214 (1958).
80. — Zh. Obshch. Khim. **31**, 474 (1961).
81. — Tokunova, I. G., Danilov, S. N.: Zh. Obshch. Khim. **36**, 1303 (1963).
82. Sykes, P.: A guidebook to mechanism in organic chemistry. London 1967.
83. Laletin, A. I., Gal'braikh, L. S., Rogovin, Z. A.: Vysokomolekul. Soedin. B **9**, 857 (1967).

Received October 11, 1973

**Springer-Verlag
Berlin Heidelberg New York**
München Johannesburg London
Madrid New Delhi
Paris Rio de Janeiro Sydney Tokyo Utrecht Wien

POLYMER CHEMISTRY

By Professor B. Vollmert,
Polymer Institute, University of Karlsruhe, Germany
Translated from the German by E. H. Immergut, New York
With 630 figures. XVII, 652 pages. 1973
Cloth DM 72.—; US $27.80. ISBN 3-540-05631-9
Prices are subject to change without notice

This book gives a comprehensive coverage of the synthesis of polymers and their reactions, structure, and properties. The treatment of the reactions used in the preparation of macromolecules and in their transformation into cross-linked materials is particularly detailed and complete. The book also gives an up-to-date presentation of other important topics, such as enzymatic and protein synthesis, solution properties of macromolecules, polymer in the solid state. The content and presentation of Professor Vollmert's book is more encompassing than most existing treatises, and its numerous figures and tables convey a wealth of data, never, however, at the expense of intellectual clarity or educational value.
The presentation is mainly on a fundamental and general level and yet the reader—student or professional—is gradually and almost casually introduced to all important natural and synthetic polymers. Complicated phenomena are explained with the aid of the simplest available examples and models in order to ensure complete understanding. However, the reader is also encouraged to think for himself and even to criticize the author's point of view. All of the chapters have been revised and enlarged from the German edition, and many of the sections are entirely new.

Contents: Introduction. — Structural Principles. — Synthesis and Reactions of Macromolecular Compounds. — The Properties of the Individual Macromolecule. — States of Macromolecular Aggregation.

Springer-Verlag
Berlin Heidelberg New York

ORGANOMETALLIC COMPOUNDS

Methods of Synthesis, Physical Constants, and Chemical Reactions
Covering the literature from 1937 to 1964
3 Vols. Second edition

Vol. 1: Compounds of Transition Metals
Edited by M. Dub. XVIII, 828 pages. 1966
Cloth DM 108,—; US $41.60
ISBN 3-540-03632-6

First Supplement. By K. Bauer, G. Haller
In preparation

Vol. 2: Compounds of Germanium, Tin, and Lead Including Biological Activity and Commercial Application
Edited by R. W. Weiss. XX, 697 pages. 1967
Cloth DM 108,—; US $41.60
ISBN 3-540-03948-1

First Supplement. By R. W. Weiss
Covering the literature from 1965 to 1968
XXV, 1116 pages. 1973. Cloth DM 112,90; US $43.50
ISBN 3-540-06304-8

Vol. 3: Compounds of Arsenic, Antimony and Bismuth
Edited by M. Dub. XX, 925 pages. 1968
Cloth DM 108,—; US $41.60
ISBN 3-540-04296-2

First Supplement. Edited by M. Dub
Covering the literature from 1965 to 1968
XXI, 613 pages. 1972. Cloth DM 78,20; US $30.10
ISBN 3-540-05845-1

Formula Index to Vols. 1-3
Edited by M. Dub, R. W. Weiss
VII, 343 pages. 1969. Cloth DM 72,—; US $27.80
ISBN 3-540-04985-1

Prices are subject to change without notice